# Mysteries Of The Multiverse

## 25 True Stories From Time And Space

### TERRENCE AYM

# MYSTERIES OF THE MULTIVERSE

Explore the Mysteries of the Multiverse and you'll discover:

- The horror that drove a British sailor mad...
- The amazing truth about an American base on the Moon...
- The astonishing story of America's war against a race of giants...
- How the genius Nikola Tesla almost destroyed the world...
- All about a deadly time vortex erupting in Antarctica...
- Why a Russian engineer claims he has a working time machine...
- Thomas Edison's fantastic attempt to speak with ghosts...
- The evidence for terrifying, giant five foot spiders in the Congo...
- A Russian spy trawler's eerie encounter with a deadly ghost ship...

...and 16 more otherworldly reports that truly are stranger than fiction!

MYSTERIES OF THE MULTIVERSE

Copyright © AYM Communications, 2009 - 2011. All rights reserved.

Published by Before It's News, Inc. - www.beforeitsnews.com

ISBN: 978-1-4675-0317-4
ISBN-13: 978-1468000856

Portions of this book were previously published.

Versions of Time Travelers Footprints Found; Scientist Builds Working Time Machine; Deadly Time Vortex Appears Over Antarctica; The Secret American Base On The Moon; Menacing UFO Fleets Circumnavigating Globe; Astrophysicist Claims Giant, Alien Spaceships Orbit Mars; Did The Nazi's Reach The Moon in 1945?; Doomsday: Earth's Core Spinning Out Of Control?, and Terrifying Giant 5-foot Spiders Spotted In Congo appeared in Before It's News, LLC.

Versions of The Mysterious Prehistoric Ropen of Papua New Guinea; The Ahool: The Monster Mega-bat That Roams The Skies of Java; Man-eating Giants Discovered In Nevada Cave; How Nikola Tesla Almost Destroyed The World; Thomas Edison's Conversations With Ghosts; The Chilling Mystery Of The Five Indian Maidens; Tragedy Of The Wailing Watcher Of The Waves; The Howling Horror In Harrison's Cave; Sadistic Surgeons From The Skies; The Mysterious Monster Of The Amazon, and Legendary Ghost Camels Of The American West, Appendix I. The Quantum Brain, Parallel Worlds, and Timeshifts, Appendix II. Quantum Entanglement And The 'Time Barrier' appeared in Helium.com.

Cover photo: Silly Little Man via Flickr

MYSTERIES OF THE MULTIVERSE

## DEDICATION

To Drs. VanderJagt and Burlet, this book is yours as much as mine.

MYSTERIES OF THE MULTIVERSE

# CONTENTS

*"The Earth is a farm. We are someone else's property."*
— Charles Hoy Fort

Forward     8

**Stranger Than Time**

1   Time Traveler's Footprints Found    10

2   Scientist Builds Working Time Machine    15

3   Deadly Time Vortex Appears Over Antarctica    19

4   Timeshift: The Monster In The Mountains    23

5   More Timeshifts: Animals That Shouldn't Be    27

6   Ships That Refuse To Die    29

7   Laser Death: 6,000 BCE    33

**Stranger Than Space**

8   The Secret American Base On The Moon    36

9   Menacing UFO Fleets Circumnavigating Globe    39

10   Astrophysicist Claims Giant, Alien Spaceships Orbit Mars    41

11   A Famous Astronaut's Last Ghostly Flight    44

12   Did The Nazi's Really Reach The Moon In 1945?    47

**Stranger Than Fiction**

| | | |
|---|---|---|
| 13 | Doomsday: Earth's Core Spinning Out Of Control? | 52 |
| 14 | The Mysterious Prehistoric Ropen Of Papua New Guinea | 55 |
| 15 | The Ahool: The Monster Mega-Bat That Roams The Skies Of Java | 57 |
| 16 | Man-Eating Giants Discovered In Nevada Cave | 59 |
| 17 | How Nikola Tesla Almost Destroyed The World | 62 |
| 18 | Thomas Edison's Conversations With Ghosts | 68 |
| 19 | The Chilling Mystery Of The Five Indian Maidens | 71 |
| 20 | Tragedy Of The Wailing Watcher Of The Waves | 75 |
| 21 | The Howling Horror In Harrison's Cave | 77 |
| 22 | Sadistic Surgeons From The Skies | 79 |

**Stranger Than Science**

| | | |
|---|---|---|
| 23 | Terrifying Giant Five Foot Spiders Spotted In Congo | 83 |
| 24 | The Mysterious Monster Of The Amazon | 87 |
| 25 | Legendary Ghost Camels Of The American West | 89 |
| | Appendices | 93 |

# MYSTERIES OF THE MULTIVERSE

MYSTERIES OF THE MULTIVERSE

# FORWARD

"There are more things in heaven and earth, Horatio, than are dreamt of in your philosophy." ~ William Shakespeare, Hamlet. Act I, Scene V.

**Forward**
There exists a world just below the comfort level of what we think is real. It's a world of half-remembered nightmares and shadows slipping through the night—a place few want to dwell in for too long a time. Yet that world exists every bit as much as the everyday one we work and play in and it may well be the hidden foundation of all we choose to call real...It's the world of the multiverse.

The world of the multiverse harbors a mosaic of mysteries—mysteries of time and mind.

Now incredible evidence is mounting that our past is not what we think it is, our future not what we expect. Time may have been manipulated and is perhaps being manipulated now. The science of quantum entanglements is revealing age-old secrets too terrible for some to contemplate.

Are the forces moving time entirely natural, or are other intelligences bending time, steering it to meet their own desires?

Mysteries of the Multiverse explores the reasons why impossible artifacts exist; reveals the myths that hide truths; delves into the secrets of time anomalies and time travel; searches for lost cities on other worlds; and confronts phantoms head on.

Enigmas have haunted humans for millennia. Twenty-five of them are presented within this book. Come, open your mind and shake off your fears—embrace these mysteries and learn of secrets that few humans have known until now...
        Terrence Aym, October 2011

MYSTERIES OF THE MULTIVERSE

# STRANGER THAN TIME

MYSTERIES OF THE MULTIVERSE

# 1. Time Travelers Footprints Found

The probability of time travel existing is high, and once time travel is achieved then it exists throughout time. Scientists such as physicists Stephen Hawking and Michio Kaku believe time travel can be achieved and may even be inevitable. Professor Ronald Mallett predicts it will become a reality before the end of this century. Kaku has gone on record. The famous physicist, quoting from T.H. White, declared: "'Everything not forbidden is compulsory.' In other words, if there isn't a basic principle of physics forbidding time travel, then time travel is necessarily a physical possibility."

Knowing that traveling through time is possible and that the universe is already billions of years old, it's conceivable that some other civilization has already manipulated time and perhaps even found ways to travel though it. It's even quite possible that humans someday create ways to travel backwards and forwards in time. If that's true, then time travel already exists.

**Inconvenient Artifacts**

Impossible artifacts litter many of the abandoned, forgotten basements of the world's greatest natural history museums. In the dim, dusty corners—stuck away with the other embarrassments of archaeology—lie some of the keys to the greatest mysteries of Mankind.

Strange and unlikely artifacts found, studied and discarded are more common than most people know. Thousands of things have been discovered that argue against the natural order that scientists have deemed as the official record of the rise of humanity.

Among the artifacts gaped at by amazed scientists and museum curators are aluminum alloy screws dated as being 100 million years old. More amazing yet are the thousands of machine engraved, manufactured spheres found in South Africa. The artifacts are estimated to have been made 2.8

billion years ago. And the incredible list goes on and on and on...

Author and researcher Michael Cremo has written several books cataloging anomalous discoveries and documenting them with eyewitness accounts, original newspaper articles and unaltered photographs. Many of the "impossible" artifacts discovered over the past 200 years still exist and are gathering dust in the basements of some of the world's most prestigious natural history museums.

### Lost Super-Civilizations, Ancient Astronauts...Or Time Travelers?

What is one to make of a machined screw found embedded in a lump of coal estimated to be 60 million years old? How did the screw get there? Who dropped it in a swamp bed that became a coal bed tens of millions of years later?

Three possible answers:

1. Non-humans visiting the planet accidentally dropped it, or,
2. The science of the history and origins of Mankind is completely wrong and the human race existed many tens or hundreds of millions of years ago and had very advanced societies that all collapsed into barbarism, or,
3. Time travelers from the future surveying the prehistoric past accidentally lost a screw. Sixty million years later coal miners discovered it.

One could make an argument that the discovery of the screw was an elaborate hoax, yet that would fly in the face of the actual documented facts. And the screw is not the only anomalous object discovered in "impossible" places. Bells, jewelry, machined alloys, remnants of unknown languages dug up in rock quarries, these and other artifacts have been tripped over by workmen, miners and excavators for hundreds of years.

With so many artifacts, fossils and discoveries made that are obviously out of place with the time line accepted by orthodox science, the objects lend credence to the hypothesis that the existence of these objects is due to time travelers that unintentionally left traces of their passage.

Time travelers win out simply by the process of elimination. The footprints that span the eons are all definitely human, especially the barefoot prints. Virtually all exobiologists agree that real visitors from the stars will be truly alien in physical characteristics and appearance. Therefore, although some may be humanoid and bi-pedal, it is against all odds that ancient astronauts exploring the Earth millions of years ago would have left human-like traces behind.

As for the idea that super-civilizations account for the ancient artifacts? Well, while it's true that evidence exists—both physical and written—that one or more truly advanced human civilizations may have preceded ours, those civilizations would have existed 25,000 to 100,000 years ago. Anthropological evidence doesn't support any human civilization existing several millions or several tens of millions of years ago. None certainly existed 260 million years ago.

The bootprints, sandal prints and barefoot prints can only be accounted for by members of time expeditions exploring the far reaches of the distant past.

## Time Traveler's Shoe Print Fossilized In Rock

Beyond impossible artifacts, the best evidence for the existence of time travel and time travelers are the human footprints left behind—from a time when almost everyone can agree no humans could have existed naturally.

James Snyder lives at the base of a mountain in the Cleveland National Forest. Snyder lived a relatively uneventful life until, in 2002, he discovered a fossilized footprint on the mountain above his property. The print had been embedded in rock estimated to be about one billion years old.

"I go out of my way to make a slip trail where nobody else has been and I was actually looking for gold," Snyder explained about his accidental discovery of the time traveler's shoe print.

## Other Ancient Tracks Left Behind By Other Time Travelers

H.W. Harkness, a mycologist, is quoted by the Proceedings of the California Academy of Science from August 1882. The proceedings describe Harkness's discovery near Carson City, Nevada and mentions that he found six sets of human tracks, each with from one to 17 footprints of shoe-soled feet.

According to George Davidson in his "The Carson Fossil Footprints" [1883] "The tracks were apparently made by more than one person."

Over the years many experienced trackers noted the fossilized tracks. All agreed the footprints were made by humans.

The stride was close to that of a normal-sized man, approximately two feet—sometimes as much as three feet.

The professor believed the tracks were made during the Pliocene period. He estimated the date to be as far back as 1.8 million years.

Dr. Harkness remarked, "If the impressions were those of any unshod animal, be it mammalian, biped, quadruped, or bird, they might differ in size, but would all be of the same pattern, which is not the case. Such a difference in shape becomes, however, quite intelligible if we suppose the footprints [were] made by men."

## Footprint From The Beginning Of The Permian Period About 300 Million Years Ago

Amazing footprints discovered in Kentucky were reported January 20, 1938 by the New York Times from an Associated Press story datelined Berea, Kentucky:

"Discovery of footprints in sandstone, so human in appearance that they might have been made by one of the earliest ancestors of man, was announced here today by Dr. Wilbur G. Burroughs, head of the Department of Geology at Berea College.

'The tracks, ten in all,' Dr. Burroughs said, 'are about 150 feet above the bottom of the Pottsville formation of the upper carboniferous system.'

"The Upper Carboniferous—or Pennsylvanian—spanned from 310 to 290 million years ago, the beginning of the Permian Period."

One paleontologist described a print found in Triassic rock. It appeared to be the fossilized leather sole of a shoe, about size 13, which showed a double line of sewed stitches, one line close to the outside edge and the other parallel at a distance of about a third of an inch. The edges of the sole were rounded off smoothly as if cut, and the right side of the heel seems to be more worn than the left [Victoria Institute, 1948].

The State Geologist of Kentucky performed extensive tests on footprints found near Berea. The prints were discovered when the overburden from a sandstone formation was removed in logging operations about 1930. One series of prints found included some arranged in a normal walking stride. Microscopic studies showed that the grain counts were greater in the soles than in the adjacent sandstone, showing greater compression within the print areas.

A shoe print was discovered in a coal seam in Fisher Canyon, Pershing County, Nevada. The imprint of the sole is so clear that traces of sewed thread are visible. The age of the coal is estimated to be older than 15 million years.

## MYSTERIES OF THE MULTIVERSE

### New Mexico Footprints 290 Million Years Old

Paleontologist Jerry MacDonald found a wide variety of fossilized tracks in 1987. The ancient tracks were made by a variety of animals and birds, located in a Permian strata running through southern New Mexico.

Among the various fossilized tracks MacDonald discovered undeniable prints of a naked human foot impossibly located in the Permian strata.

The Permian strata dates from 290 to 248 million years ago—many millions of years before animals, birds, dinosaurs, and man existed.

In July 1992, the Smithsonian Magazine ran an article on MacDonald's tracks, "Petrified Footprints: A Puzzling Parade of Permian Beasts."

Smithsonian admitted the mystery and acknowledged "what paleontologists like to call, 'problematica.'" It described what appeared to be human footprints. Humans "evolved long after the Permian period, yet these tracks are clearly Permian."

### Does Time Travel Really Exist?

The probability of time travel existing is unknown; the possibilities are endless. As has been pointed out, if time travel is ever accomplished then it must necessarily exist now.

We are a young race orbiting a third generation star on the outskirts of our galaxy. To declare that time travel is impossible—or that almost anything is impossible—would be shutting the doors on reason and imagination. It would also be the height of arrogance.

In a sense we are all already time travelers, like flotsam drifting with the current of a river. When we learn how to swim upstream and downstream then we will have the answer to the question.

And perhaps someday it will be our children's children that step into the past of 260 million years ago and leave a fossilized record of their adventure for all to see.

## 2. Scientist Builds Working Time Machine

What person hasn't dreamed at least once of returning briefly to a favorite era in the past or seeing a real glimpse of a fantastic future hundreds, even thousands of years from today?

Despite a spate of articles claiming that travel to the past is flatly impossible, physicists like Stephen Hawking and others say it's not.

And now an engineer named Vadim Alexandrovich Chernobrov claims to have accomplished what others say cannot be done—he's built a working time machine.

At least that is his claim. And he says he's tested it and it works. But there are some limitations with its capabilities.

**Potential Dangers**

Chernobrov cautions about potential dangers. "Those who attempt to return to the past to keep certain historical events from occurring—for example, the collapse of the Soviet Union—would fail in the attempt and would run the risk of not being able to return to the future."

The engineer's amazing time machine is a metal sphere that looks much like some of the early Soviet and American satellite spacecraft. Outside it has a diameter of two meters and inside is housed a meter wide space that can accomodate a time-traveling chrononaut.

The device achieves travel through time, according to its inventor, by utilizing strong magnetic fields that resonate about the capsule and affect the natural flow of time and causality by literally increasing or slowing the flow of time.

During experiments, Chernobrov has shown that very precise chronometers, synchronized before the field is energized, experience a significant divergence in time after one has been placed inside the time machine while the other is kept far from the energized field as a control.

The chronometers reveal time is definitely warped within the machine's magnetic fields.

## First Time Machine Made At Moscow Aircraft University

According to a chapter translated by Lena Ksandinova, from Chernobrov's book "Time Travel" [Armada, 2002 2nd edition], "The first Russian time machine was constructed by the scientists of Moscow Aircraft University.

"To understand, how the machine impacts live creatures, Vadim Chernobrov, an experiment administrator, put cockroaches inside. The insects died. They modernized the machine and put mice inside. The effect on them was the same. Scientists realized what the cause of death was: on different parts of the body time ran with a different speed. They fixed the machine once again and put a puppy called 'Lunar Rover' inside. He was inside for 108 minutes. The puppy survived, but behaved like he was having rabies.

"Then experiments on people came. Nine scientists, among which was Chernobrov, took part in it. Instruments inside the machine recorded a 3 percent time shift, and changes in machine size. Experiment participants, however, felt nothing but moderate arrhythmia and colored circles in eyes."

The experiment with people was the most dangerous. Some scientists familiar with the experiment condemned it as reckless.

"We came to conclusion that time is 3–dimensional," explained Chernobrov. "So in fact we can move in time backwards and forwards."

## Details On Larger Model

As Ksandinova reported, a larger model of the time machine was built to carry humans. It was constructed in a remote forest in Russia's Volgograd region. While the device had a low power capacity it still affected the flow of time by three per cent; that temporal permutation was measured with symmetrical crystal oscillators.

At first, the researchers spent periods of five, ten and later, twenty minutes in the field of the time machine; the longest stay recorded lasted about half an hour. Chernobrov reported the volunteers felt as if they had shifted into a different world: "they felt life here and there at the same time as if some space was unfolding."

## Is Time Being Manipulated?

If time travel has, does or ever will exist then by its very nature time travel always exists. "What is troubling is that we have no way of knowing that they aren't already doing it. For those of us in the daily stream of world events, a shift in history might wipe out thousands of people and change entire governments. But for us, the change would go into our memory of events as they happened during our lives. That good friend we went bowling with last night might disappear before the next morning and we would not notice his loss. By the time we awake, the person never existed."

## Notes From A Pioneer

In his notes on the early time machine experiments, Chernobrov wrote:
"In experiments on deceleration and acceleration of a physical Time in small closed volume conducted since 1988, among other things the effect electromagnetic of fields on space-time continuum was checked.

"The experimental system for such effect represented set of electromagnets, are connected among themselves in serial and in parallel and installed on the globe-shaped surfaces. In various experiments from 3 up to 5 such surfaces, Electromagnetic Working Surfaces (EWS) was used. All EWS layers of various diameters were installed consistently in each other.

"The maximum EWS size was about meter, the minimum EWS diameter (internal) was equal to 115 mm, that has appeared sufficient to place inside of EWS the gauges of the control and experimental animals (various kinds insect and laboratory mouses), to investigate the effects of converging spherical electromagnetic of waves."

Presumably you can reach Vadim Chernobrov through his website if he isn't off somewhere traipsing around in time.

Chernobrov's complete time machine report, "Experiments on the change of the direction and rate of time motion," can be found here.

## 3. Deadly Time Vortex Appears Over Antarctica

Disturbing news has been leaking out from the giant continent at the bottom of the world. Some scientists manning lonely outposts under the drifting and shifting aurora are nearly paralyzed with fear. Their clipped reports are being reviewed by astonished superiors back in the home countries.

Russian researchers posted near the giant South Pole sub-glacial Lake Vostok have reported eerie anomalies and incidents over the past few years that sometimes seem to border on the frayed edge of creeping madness.

**Artificial Structure Found Under Two Miles Of Ice?**

During April 2001 one of the world's great secrets was revealed: an ancient structure or apparatus that lay encased miles under the hard Antarctic ice was detected by a roving spy satellite. The US military immediately moved to quash the reports and the mainstream news media dutifullycomplied.

Despite the news blackout, reports still surfaced that a secretive excavation project had commenced on the heels of the discovery. Some European countries formally protested the excavation by the US military.

**Excavation Deep Into The Ice...What's Down There?**

"If it's something the U.S. military has constructed down there, then they're violating the international Antarctic Treaty," said an aide to Nicole Fontaine, then the European Parliament's French president. "If not, then it's something that's at least 12,000 years old, which is how long ice has covered Antarctica. That would make it the oldest man-made structure on the planet. The Pentagon should heed the calls of Congress and release whatever it's hiding."

The federal government and the Pentagon ignored the calls.

# MYSTERIES OF THE MULTIVERSE

## High Technology And High Strangeness

Soon after, some military observers noted that robotic devices were being shipped to the South Pole and speculation erupted about the belief by some that the US Air Force had transported their mammoth nuclear-powered tunnel boring machine, the Subterrene, on a C5-A to a secret Antarctic base.

The Subterrene—a cylindrical vessel that is said to be manned with a crew of four to six—is capable of subterranean travel and has undergone tests in Nevada, Colorado and New Mexico. Designed to bore through hard rock strata, drilling into the ice would be like a laser cutting through a marshmallow for the Subterrene.

Following the frenzied events of early 2001, the news broke of a mysterious medical emergency that forced an evacuation of unnamed personnel during the depths of the Antarctic winter–the first event of its kind during the dangerous South Pole winter season.

Shortly after that the region was shaken by an unusual earthquake. Seismologists located the temblor at the epicenter of the buried structure in East Antarctica.

Still the military resisted making any comment.

A magnetic anomaly formed, intensified, and spread to the vicinity of the Russian Vostok base. Russian researchers were shocked and puzzled.

Meanwhile, the American military airfield buzzed with activity as flights came and went at a dizzying pace. Heavy machinery—some pretty exotic—appeared on the bleak Antarctic ice sheet. Unverified reports claimed that the nuclear-powered earth borer Subterrene arrived.

Finally, the American and European media were pressing hard for some believable answers when 9/11 occurred and the US was suddenly under terrorist attack. Domestic and international focus immediately shifted from the Antarctic to New York City and Washington, D.C.

The mysterious events in Antarctica were forgotten.

## 2002: A TV Crew Disappears?

A California TV crew filming in the Antarctic reportedly went missing in November 2002.

Supposedly, a video discovered among the crews' personal effects by a special U.S. Navy SEAL rescue team tasked to find the filmmakers confirmed earlier reports of a huge artifact buried under the ice—a prehistoric machine that may be of alien origin.

"The U.S. government said it will seek to block the airing of a video found by Navy rescuers in Antarctica that purportedly reveals that a massive archaeological dig is underway two miles beneath the ice," stated a press release that appeared briefly on the studio production's website.

Later, the missing TV crew story was said to be a publicity stunt.

## Time Vortex Erupts

Yet, as one bizarre event followed another, a research team of US scientists from McMurdo and UK scientists from Halley accidentally came across a mind-numbing discovery. While working on a joint weather research project on the ice field between McMurdo and the giant Russian Vostok base, the team witnessed the creation of a spinning vortex of time.

US physicist Mariann McLein allegedly testified that she and her colleagues became aware of a "spinning gray fog" in the sky over their heads. They initially dismissed the phenomenon as merely part of a random polar storm.

The spiraling vortex, however, did not disperse. Stranger still, despite gusts of wind and briskly moving clouds overhead, the weird spinning gray fog remained stationary.

## Intense Magnetic Anomaly Appears

Deciding to explore the odd phenomenon, the group took one of its weather balloons and attached a meteorological instrument to it that calibrated temperature, barometric pressure, humidity, wind speed as well as a scientific chronometer to record the times of the readings.

After attaching a cable to the balloon, and securing the other end to a winch, they released it. The balloon and instrument package soared upwards and were immediately sucked into the gaping maw of the swirling vortex.

The balloon and instruments disappeared.

## Tunnel To Past Emerges

After several minutes, they decided to retrieve the balloon. Despite some difficulty with the winch they succeeded in bringing the balloon back to earth and checked the instruments. McLein stated that everyone was stunned by the readout on the chronometer. It displayed a date decades in the past: January 27, 1965.

McLein claimed the experiment was repeated several times with the same result.

Later, she said, the entire episode was reported to military intelligence and passed on to the White House. The strange vortex phenomenon–a highly magnetic tunnel to the past–was reportedly code named The Time Gate by military intelligence.

**Magnetic Time Tunnel To The Past?**

As the phenomenon was generated in the same general vicinity as the discovery of the giant apparatus deep under the ice, it's thought the two may be related.

If indeed a magnetic time vortex is appearing and disappearing over Antarctica–and if the phenomena is not natural, but generated by some unknown technology deep under the icecap–it may reveal the physics of time and could potentially allow control of the past, and by implication the future.

If true, it's no wonder the military is so intensely interested and so closed-mouthed about it.

## 4. Timeshift: The Monster In The Mountains

The sleepy hamlet of Argentiera, nestled between mountains on a ridge near Sardinia, Italy, has been disturbed very few times since its founding during the Roman Empire.

The town sprang into existence due to the discovery of silver. Mining of that precious metal continued there for centuries.

The people of Argentiera were always quiet and practical. Until March 1947 only the battles of World War II and several moderate earthquakes caused any sort of stir.

Yet the morning hours of March 3, 1947 became engrained in many residents' memories , for that was the day the monster appeared from the mountains.

[Versions of this incredible episode have appeared over the years. Ostensibly, the first report of the events that unfolded in the tiny Italian town appeared in an obscure local newspaper during the late 1940s—a paper long since defunct.

Another version of the incident surfaced in an early 1950s account written by an Austrian journalist. Later that was translated into a sensationalized account by a bi-weekly tabloid published at the time in Buenos Aires, Argentina.

Part of the witness descriptions of the events have been reconstructed from translations of excerpts from the original Italian story. Because more than 60 years have passed, and detailed documentation of all the facts have been obfuscated by various versions, the accuracy of some of the witnesses' testimony cannot be assured. The story itself, however, is so amazing it deserves being retold.]

Father Palo Vinecelli, a devout and educated Roman Catholic priest was taking his usual morning stroll from the church rectory to the village's main square. Signor Gesseppi Rivoletti, the priest's portly friend who served as both mayor and town baker, accompanied Vinecelli on the walk.

As the two men walked they conversed about an upcoming marriage to be officiated over by the priest. Rivoletti had been commissioned to make a four-tired wedding cake for the reception.

Father Vinecelli later recalled the mayoral baker had been wondering aloud about the frosting on the cake when their attention was drawn to a growing rumble coming from the vicinity of the nearby mine.

"I was at a loss as to the origin of the sound," Vinecelli later told reporters. "My friend, Gesseppi at first thought it to be a tremor. We do have mild earthquakes in the region. Myself, I believed that some disaster was occurring in the mine. A cave-in, perhaps. To me the sound was emanating from the mine shafts.

"But soon I saw two army helicopters approaching our village. I was surprised as helicopters are not a common sight. I only saw one during the war. They came around a nearby hill and flew just above the tree tops.

"Then we saw an armored column approaching...soldiers driving vehicles like I hadn't seen since the war. I remember all the noise and dust in the air."

Rivoletti turned and remarked how strange it was when a military plane appeared out of nowhere and flew right over the valley. That was a sight!

As the events unfolded before their eyes, the two shocked men stood on the side of the road. Soon other villagers joined them.

Vinecelli recalls the soldiers appeared tense. Officers shouted orders at the men. But the machines they drove were so noisy most of the their orders were lost in the cacophony. What the Father remembered was an atmosphere of controlled chaos and palpable fear. It hung over the scene like a canvas tarpaulin.

Father Vinecelli: "Suddenly we sensed a change in the atmosphere. The air seemed to crystallize around us. And then it began wavering and shimmering like heat rising from a tar road on a hot summer day.

"A cool, fresh breeze had been blowing in that morning, carrying with it the smell of the trees and new grass. That abruptly changed. Somehow the air became heavy. A putrid smell accompanied sudden gusts of hot, moist air. The wind reeked of steamy jungle growth—certainly something one would never encounter in Italy! And Gesseppi smelled it too.

"It was so overpowering I thought I might gag."

From out of the noise and dust, the confusion and foul wind came the thunderous roar of some great beast—a roar not heard upon the Earth in more than 65 million years.

The priest described the animal as being more than two stories tall. Its hide was a mottled gray-green color. The body seemed completely hairless.

It stood upright and continuously thrashed its small arms. The arms were puny compared to its legs. Vinecelli described them as being the legs of Satan.

Rivoletti later recalled his impression of the creature's legs and called them "giant, muscular, mud-splattered pillars."

Father Vinecelli: "The beast had a long, thick tail that it seemed to use for balance and as a defensive weapon. My God, that tail! It swung back and forth through the air like a whip. Just like a whip. And it used it to slash at the road and at soldiers approaching it from behind. Yes, it was just like a whip...wielded by the Devil himself.

"The thing had a huge head with evil-looking, deeply sunken eyes. And it had huge jaws measuring at least six feet."

What Father Vinecelli described is a perfect description of a Tyrannosaurus Rex, the king of the thunder lizards.

A T-Rex prowling along a 20th Century Italian highway?

Another gust of the strange wind blew up as the creature hesitatingly advanced towards the village.

Father Vinecelli: "The soldiers brought their weapons to bear on the beast. The men looked scared, but I didn't blame them. We were all scared. Most of the villagers had run away, but a few, including Gesseppi stayed back along the wall of a building and continued to watch the amazing events.

"I heard rifle fire. Some of the rounds hit the roadway and others ricocheted off rocks. None of us knew if the great beast had been hit, but we all heard it bellow.

"It seemed confused. I think the beast was itself scared."

When the soldiers fired at it, the T-Rex turned and dashed off with surprising speed back across the road towards the shelter of a copse farther up the hillside.

As Father Vinecelli and the others watched in utter amazement, the soldiers gave chase. The helicopters followed.

But after reconnoitering the area for almost 30 minutes the soldiers called off the search. Somehow they'd lost track of the creature.

Or had they?

Had some time gate closed? Was the dinosaur swallowed up in the same manner it had been briefly thrust into the 20th Century?

And how long has it been wandering around the Italian countryside before the military was called in to confront it? Father Vinecelli never got a straight answer.

The Tyrannosaurus was never seen again, much to the relief of the good citizens of Argentiera and, no doubt, the Italian militia.

Monsters out of time, however, are not only encountered in Italy. Numerous sightings of creatures out of time have been reported throughout the centuries. We'll explore some of those sightings in the next chapter.

## 5. More Timeshifts: Animals That Shouldn't Be

Timeshifts and a monster in Italy...what other impossible creatures roam the labyrinth of space and time?
Plenty.
Through the ages the reports are too numerous to catalog. The gamut of creatures that are out of place and out of time seems like an animal menagerie tended by mad zookeepers.
There exists a world just below the comfort level of what we think is real. It's a world of half-remembered nightmares and shadows slipping through the night—a place few want to dwell in for too long a time.
Yet that world seems just as real as the everyday one we work and play in, it may well be the hidden foundation of all we choose to call real.

### Pterosaurs, Pterodactyls And Thunderbirds

The natives of North America witnessed giant birds and other flying monsters in the skies long before the arrival of Europeans. Many tribes myths and legends stretch back through the millennia. From the Northwest to the Southeast, from the Atlantic to the Pacific, encounters with terrifying raptors have fueled the stories shared over many a campfire.
Those sightings continue today.
A quick search on Google will reveal countless entries of creatures resembling extinct animals spotted in the skies—classes of animals that perished eons before human s took their first halting steps on Earth.
Prehistoric pterosaurs seem to be spotted the most often and sometimes in the most unlikely places. Some of the more spectacular cases during the last century include:
1927: Sausalito, California. Enormous bat-bird with a torpedo-like head spotted in sky by dozens of witnesses.

1948: Santa Barbara, California. Giant bird spotted; no feathers; head shaped like torpedo.

1948: Alton, Freeport, and Caledonia, Illinois. Giant monster birds reported; anvil head, stinger tails. Seen by hundreds.

1961: Over Hudson River, New York. Giant bird described by witnesses as "prehistoric in shape" approaches airplane in flight.

1966: Point Pleasant area, West Virginia. Monster reptile reported by many witnesses. Creature seen soaring in sky.

Are there natural phenomenon unknown to us? How can prehistoric creatures appear without warning millions of years after they are supposed to be long dead? How can they just as suddenly vanish? A handful of cases might be dismissed, but thousands have seen such creatures. How many tens of thousands of incidents like these have occurred, but were never recorded?

Animals that inexplicably appear and disappear are not limited to prehistoric creatures. Countless cases exist of kangaroos, camels, lions, tigers, elephants, ostriches, and more spotted, chased, and lost. Yet no zoo, circus, or individual has reported any animal missing. Typically, the animal is seen for one or two days and then simply vanishes.

1886: Along the Connecticut and Hudson Rivers. "Horned monsters" reported by many.

1887: Memphis, Tennessee. Hundreds of black snakes suddenly appear in the city streets over a two block area.

1889: Lincoln, Nebraska. African lemur found "in convulsions" in back yard of a suburban home. Animal later dies.

1931: Patagonia, Argentina. Bloody beast seen by townspeople. Said to resemble extinct giant ground sloth.

1953: Springfield, Missouri. Twelve cobras and one boa constrictor found in a small area.

1961: Muskegon, Michigan. Eight-inch fish found in small puddle in middle of dry street.

1970: Newton, Kansas. Ten-foot long alligator found by surprised owner in sealed basement of suburban house.

Could it be some natural force plucks these hapless animals from their normal habitats whisking them out of space time and plopping them somewhere else? If so it is an entire field of physics that science has missed.

Curiously, instances of this phenomenon occur more frequently in some parts of the world than in others. Is this a clue that an intricate space time field exists that sometimes twists and lets things fall through?

Upon investigation it surely seems that some devilish twists in the space time continuum exist. Evidence of timeshifts occur throughout history.

The unsettled citizens of Argentiera, Italy might have found kinship with the ancient tribes people of Northern Arizona. In 1924 the Doheny Scientific Expedition was shocked to find an impossible rock carving in the Hava Supai canyon: our old friend Tyrannosaurus Rex depicted in all his glory.

Some years later another scientific expedition stood amazed before the work of a prehistoric artist. Far to the northwest of Arizona in Big Sandy River, Oregon a rock carving displayed the unmistakable image of a Stegosaurus.

No real evidence exists and no rational theory can be formulated that can place the dinosaur race and humans in the same era. At best the ancestors of humans appeared some four million years ago while all evidence shows that most dinosaurs had died out about 65 million years ago. That's quite a gap to bridge.

Yet incontrovertible evidence exists that humans saw dinosaurs and see them today. And every anthropologist can attest that prehistoric artists only recorded what they actually saw.

Only one answer remains: timeshifts.

MYSTERIES OF THE MULTIVERSE

# 6. Ships That Refuse To Die

Ghost ships and the legends they leave in their wake have sailed through the centuries. Many famous tales haunt the waters and the misty half-forgotten docks of less traveled ports that only lonely sailors haunt.

The list of famous ships that haunt the seven seas could fill books—and sometimes they have.

Perhaps the most famous is the legendary Flying Dutchman, a cursed ship that was said to appear before certain disaster struck mariners.

One of the most eerie and inexplicable ghost ship encounters, however, did not happen centuries ago, nor was it likely ghostly. The event took place in 1967. It had all the earmarks of a timeshift.

Several versions of this story have been related during the past quarter century, among them a version by a less than credible tabloid.

The true incident is remarkable and hints at properties of physics and time that scientists have barely begun to unravel with their studies of quantum entanglement.

The nightmare began aboard a Russian trawler that lay anchored not far off the southern coast of Cuba. The trawler was a spy ship and had been steaming up and down off the Eastern coastline of the United States until the Coast Guard chased it off.

On May 18, 1967 the trawler dropped anchor and remained there for several days. Aboard were fisherman who actually fished as part of the Russian ruse, two electrical engineers that were assigned to test and evaluate new eavesdropping equipment, a KGB political officer, and of course the captain and first mate.

The events that transpired on May 21 partially leaked out before the Soviet government put a tight lid on it. The story that circulated later had some truth, but was enhanced with conjecture and exaggeration.

After the collapse of the Soviet Union more of the story emerged and was carried by some of the Russian news media during the mid-1990s.

Interviewed after the event, both electrical engineers declared that what they had witnessed "...defied explanation under scientific principles currently understood."

That quote managed to appear in many of the early versions of the incident that made it into the fringe press.

Much of the rest of the event, however was distorted by the media that bothered to report it at all.

Here then is the recollection of the events that occurred according to the two Russian electrical engineers:

Galai and Turchuv had been assigned to the trawler almost at the last minute. As specialists, they were qualified to rate and assess the performance of the electronic eavesdropping equipment, but neither man had much experience on seagoing vessels. A Russian destroyer had transported them to the Atlantic where they rendezvoused with the trawler about 400 nautical miles due east of South Carolina.

After the trawler was shooed away by the U.S. Coast Guard, the ship made port in Havana. It remained there for two days until steaming to the south and dropping anchor in a shallow part of the Caribbean.

Well within the region called the Bermuda Triangle.

Everything seemed normal until the morning of the third day they lay at anchor. Turchuv was the first to notice the electrical equipment malfunctioning.

First the navigation, radio and sensing equipment went down followed by malfunctions with the ship's radar and sonar. Anomalous electrical fluctuations plagued the vessel.

The malfunctions were bothersome and frustrating yet only a precursor to the bizarre events that followed.

Galai recalled, "The electrical problems increased. We had our hands full trying to hold everything together. It was as if all the circuits had suddenly gone crazy. And then a seaman rushed in—I believe Turchuv and I were in the radio room at the time—and I told him to leave, couldn't he see we had a disaster on our hands?

"But the man insisted we accompany him to the deck and informed us the captain had requested it. So we left the equipment there and followed him to the deck.

"What greeted us when we arrived on the deck was like something from another world. Everyone was staring and pointing off the starboard side of the ship. In the distance I saw a wooden, three-masted clipper ship. Actually, I thought it must be a mirage, if one can see a mirage at sea.

"She was like something straight from the history books—like something from 150 years ago. She seemed to be listing and we could see people on her decks.

"Although it was a clear day, I noticed a fine green mist near the vessel. The mist obscured some of the details, but we could clearly see the people and the masts and the sails. We also heard shouts that we all took as cries for help."

The trawler's captain ordered a party to lower a skiff and board the distressed ship to determine if assistance was needed. Shortly thereafter, six seamen set out for the stricken vessel.

"We all watched our men cutting through the choppy waves," Galai said. "They were nearly at the ship when the green mist became thicker and obscured everything."

"Minutes later the green mist dissipated. When it was entirely gone we discovered our men and the ship were also gone. We saw the sea and nothing more."

The captain thought the clipper must have gone down and taken his men with it. He immediately ordered a search.

"But our men and the strange ship were nowhere to be found. No debris, no bodies, no evidence remained that the mystery vessel or our men had ever existed."

Neither craft had sunk or broken up. They were just there one moment and gone the next.

Many years later a report speculated that the phantom ship and her crew were trapped in a timeshift, a warp, for at least 150 years.

A witness also speculated, based on observation, that the crew and passengers of that tragic ship may have been unaware of their plight. They might not have known what happened and the last 150 years may have seemed no more than 150 seconds to them.

And what of the six Soviet sailors who vanished along with the mystery ship?

They never returned.

## 7. Laser Death: 6,000 BCE

Footprints through time are intriguing, but perhaps more intriguing are two astounding examples of prehistoric death by high technology. Both are paleontological clues that either time travelers—or something from another world–walked our planet thousands of years in the past.

**The Auroch And The 'Death Ray'**

In a case, in a dim wing of the austere Museum of Paleontology in Moscow, sits an exhibit that should not exist: the skull of an auroch with what appears to be a bullet hole through the forehead.

Aurochs, ancient ancestors to contemporary bison, are large animals that flourished during the Neolithic period spanning from 10,700 BC to about 4,500 BC. The Moscow specimen has been dated to roughly 6,000 BC making it 8,000 years old.

During the early Neolithic Era human development had barely reached the invention of agriculture and the creation of clay pottery. Tools made out of metal did not begin to appear until about 6,500 years ago.

Yet the 8,000 year old auroch skull sits in that exhibit, a silent testament that something with technology that wouldn't appear for almost 8,000 more years put a hole in its skull and killed it.

Aware of the anachronism, Soviet scientists subjected the skull to forensic analysis almost 60 years ago.

When the creature died, humans only possessed crude stone axes. The auroch, they speculated, must have been killed by a bullet. The wound in the skull exhibited the same cranial damage as that inflicted by a high-powered, larger caliber bullet such as that fired from a rifle. A slower speed would have resulted in more extensive damage to the bone mass. Even if the animal received a terrific blow from a sharply chiseled stone ax parts of the skull would have shattered.

The consensus of the forensic experts was the beast had been felled by a bullet—or something like a bullet.

And there the matter rested until the 1980s when the the investigation reopened. This time a new team studied the edge of the hole using a microscope. What they found under closer examination created a bigger enigma that the first team encountered.

Traces of burn marks on the bone and telltale fusing was evident. The scientific analysis left little doubt. No bullet had felled this ancient auroch.

The creature had been shot at and killed by a high-intensity laser beam.

**A 38,000 Year Old Murder**

About 1,500 miles from Moscow stands the imposing British Royal Museum of Natural History in London. This museum also has an odd skull. Unlike Moscow's skull London's is human.

The skull has been dated to the Paleolithic Era some 38,000 years in the past. Excavated in 1921 during a British archeological dig in Northern Rhodesia (now Zambia), the skull was cleaned, marked, studied and then shipped to the Royal Museum where it still resides.

The owner of the skull was an adult male. Most of the skeleton was also retrieved.

Although skulls from the Paleolithic Era are not too uncommon, this particular specimen is unique. A bullet hole is evident on the left side of the skull with an exit wound on the opposite side. The hole is a perfect circle about one-third of an inch across. As is typical in the case of a bullet fired from any distance except point-blank range, the hole exhibits no radial splits or fractures. It is a clean wound. The bullet entered the skull, did its damage, and blew out part of the cranium on the other side.

Forensic analysis of the fragments revealed the cranium was blown outwards from the inside as the bullet emerged. That too is typical of a high-caliber gunshot wound from an exiting bullet. Firearms experts believe the bullet must have been fired from a rifle.

Arguments have been made that the skull is from a man who lived no more than 150 years ago, but the disposition of the body and the site where it was found make that hypothesis very unlikely. The remains were discovered in a cavern located near the Zambesi River. Deep in a pit some 60 feet deep encased under layers ancient clay with deposits of lead.

MYSTERIES OF THE MULTIVERSE

# STRANGER THAN SPACE

## 8. The Secret American Base On The Moon

When American astronauts left the Moon for the final time as the Apollo 17 mission wound down, none realized then that NASA would not return. Three more missions were scheduled, yet the program was scrapped.

Why?

For almost 40 years that question has been the nexus of an intense debate amongst space enthusiasts and conspiracy theorists.

Although guesses ranged from lack of money and enthusiasm to an ominous alien threat, the truth seems to lie somewhere between the mundane and the incredible.

It cannot be denied that some years later—when America finally returned to the Moon in the form of a Department of Defense high resolution mapping satellite—more than two million photos were taken of the entire lunar surface. Of special interest: the far side of the Moon.

All the data was stamped "Top Secret" and rushed to a team of tight-lipped experts manning red-lit, windowless rooms in the secure underground chambers of the National Security Agency.

And then the decades dragged onwards under the cover of black projects funded by black budgets.

A parade of military officers, aircraft controllers and former intelligence personnel appeared before the National Press Club in Washington, D.C. During the day of September 27, 2010. Among them was ex-intelligence officer who spoke about a base located inside a shallow crater on the Moon. He claimed the codename of the base—appropriately enough—is Luna, the formal name for the Moon.

Factions within the Pentagon, specifically the United States Navy, happened upon evidence that the Moon is inhabited and that at least one base exists. Circumstantial evidence also makes some suspect that other bases exist on the far side of the Moon.

Another Disclosure Project participant, former US Air Force sergeant Karl Wolfe, testified that "I was asked to go over to this facility on Langley Air Force base where the National Security Agency (NSA) was bringing in the information from the Lunar Orbiter." [Disclosure Project]

"They took me into this laboratory, I took a look at the equipment, there was an 'Airman 2nd Class' in there, and I was an A2C as well."

Wolfe went on to testify that "We walked over to one side of the lab and [I was told], 'by the way we've discovered a base on the backside of the Moon.'"

According to Wolfe's deposition, the NSA man showed the Airman a mosaic of the lunar surface photographed by the Lunar Orbiter "...he pulled out one of these mosaics and showed this base which had geometric shapes, there were towers, spherical buildings, there were very tall towers and things that somewhat looked like radar dishes...but they were large structures."

Although the ONI and other Pentagon departments believe that the bases are not of Earth origin, the preponderance of evidence suggests they were built by humans within the past few decades and were constructed with some of the $40 trillion that's been funneled into the military and intelligence communities' black projects since the mid-1960s.

Much of the base may be under the surface and connected by ancient sub-lunar lava tubes.

### A 'Dark' Space Program?

Rumors in the aerospace industry and the US Air Force have circulated for years about a separate, secret space program. The visible program run by NASA was for public consumption while the real work—and the focus of the nation's national security interests—was done under cover.

Many still wonder why the Vandenberg military shuttle launch site was suddenly abandoned during the 1990s with little fanfare. After spending many millions building the space launch facility designed to accommodate the Space Shuttle, the USAF abruptly tore it down. The facility never launched a single shuttle.

Some at Lockheed insinuate that the fourth generation advanced Aurora platform has full space capabilities and the USAF no longer needed the shuttle, but the smart money is placed on the black platforms that many around the globe have been reporting as UFOs.

## Space Capable Military Platforms

Among the exotic space capable craft that may well be ferrying military personnel between the Earth and Lunar bases is the "TR-3b." It's not an official military designation, but one that' been "assigned" to the mystery craft by aerospace insiders that have been kept out of the need-to-know loop.

The TR-3b has been seen by witnesses leaving or arriving at some USAF bases such as Scott Air Force base in southwestern Illinois and Wright-Patterson in Ohio.

The craft is utterly silent, stealthy, and is not designed to fly through the atmosphere. Many witnesses describe the platform as gliding and swiveling when it changes vector. The frame is not an airfoil and it does not bank when it turns, it just points its nose in the new direction when it changes direction.

As the new space programs of India, Japan and China continue to ramp up more information on the mysterious lunar base may be revealed.

## 9. Menacing UFO Fleets Circumnavigating Globe

Frightened UFO investigators ask: Is the world is being invaded?

Thousands of awestruck people around the world are seeing—and sometimes recording—literal fleets of UFOs. The unworldly craft dominate the skies over major cities with complete impunity.

### Terror Over Taiwan

In the video they arrive without warning and dart across the skies like they own it...and maybe they do.

The UFO armada appeared over Taiwan to the amazement and consternation of pedestrians on the streets far below.

Silent, maneuverable, and definitely under the control of some unknown intelligence, the UFOs dipped and looped while the authorities on the ground stood helplessly gaping up at the sky along with thousands of stunned citizens.

### Massive UFO Fleet Invades Seoul, South Korea

More than 20 UFOs converged on Seoul, South Korea's largest city. According to news reports the UFOs dominated the skies over other parts of South Korea as well, yet the South Korean officials kept their aircraft grounded on the advice of an American military officer that was in satellite contact with Vandenberg Space Command in California during the incident.

His reported advice to frightened South Korean officials was direct and blunt: Don't chase after them; it's too damn dangerous.

### UFO Armada Commands Lebanese Skies

Amazed eyewitnesses swore that huge UFOs buzzed the expressway at Antelias Jounieh in the city of Beirut. They made their claims on a national TV show hosted by Il Maestro Nichan Derharoutiounian a respected journalist. The highly rated show was broadcast across the nation on the LBCI network.

Witnesses testified that hundreds of cars pulled off the roadway and stopped when more than 50 UFOs converged overhead. Some of the alien craft descended almost to ground level before rising again and rejoining the fleet.

The reports were carried across the Middle East and Europe, but as is normally the case the American news media completely ignored the furor.

**Fleets Of UFOs Now Appearing All Over The World**

China

The Chinese are picking up hundreds on radar. Meanwhile they're forced to close commercial airports as some gigantic craft appear over runways. Reports from China describe some of the alien craft as being half a mile long.

Saudi Arabia

In Saudi Arabia UFOs appeared near Riyadh and spelled out Allah's name in the sky. Thousands fell to their knees in prayer struck with awe and fear.

Mexico

Residents of Guadalajara, Mexico spied hundreds of darting, dashing UFOs zoom over their heads. The Mexican military remained helpless.

Residents of Guadalajara have seen UFOs for decades

Great Britain

Across the Atlantic in the United Kingdom, expressways like the M25 are being buzzed by UFOnauts...some Brits snapped photos of the UFO sky hordes as the craft soared overhead.

Manhattan Island, New York

New Yorkers gawked as the sky above Manhattan filled with mystery ships. The normally staid city folk yelled and pointed at the sky—their frank amazement tinged with fear.

From Brazil to Australia...from Russia to Canada...the skies are bursting with UFOs. What are they really? Where do they come from? Why are they here? What do they want? Are they harbingers of doom? Monitors of an approaching worldwide cataclysm?

Some hazard guesses, but no one really knows...

## 10. Astrophysicist Claims Giant, Alien Spaceships Orbit Mars

**A**strophysicist Dr. Iosif Samuilovich Shklovsky calculated the orbital motion of Martian satellite Phobos and came to the jaw-dropping conclusion that the moon is artificial, hollow, and basically a titanic spaceship.

The scientist is world-renown for penning the classic science book, "Intelligent Life in the Universe" with famous Cornell University professor, the late Carl Sagan of PBS and Voyager space probe fame.

**Fear And Horror**

Mars' two moons, Phobos and Deimos, translate into "fear" and "horror." As Mars is named after the god of war, the names seem appropriate. Both satellites were discovered in 1877 by U.S. astronomer Asaph Hall who never guessed they were artificial.

Both moons are extremely odd, especially the tumbling moon of fear: Phobos. Shklovsky puzzled over them.

**Deeply Troubling Facts**

Two facts deeply troubled Shklovsky.

First, both moons are too small. No other planet in the solar system has moons as tiny as the Martian moons. They're unique.

Second, their origin bothered him. Were they captured asteroids as others assumed? No, they could not be! Their orbital plane was all wrong. And they're too close to Mars. Much too close. Even more amazing—Phobos changes its speed from time to time.

Impossible, yet true!

## Phobos Is Shaped Like Interstellar, Generation Starship

Russian astronomer Dr. Cherman Struve spent months calculating the Martian moons' orbits with extreme accuracy early in the 20th Century. Yet, Shklovsky astutely noted, as the years progressed into decades the mystery moon's orbital velocity and position no longer matched its mathematically predicted position.

After lengthy study of the tidal, gravitic, and magnetic forces, Shklovsky came to the inescapable conclusion that no natural causes could account for the origins of the two odd moons or their bizarre behavior, particularly that exhibited by Phobos.

The orbit of that fantastic moon was so peculiar, so bizarre, that Phobos had to be a gigantic spaceship.

Every other possible cause was carefully considered and resignedly rejected. Alternate explanations either had no supporting proof or the math was wildly off.

So, Phobos had to be accelerating as it lost altitude, yet could the outer fringes of the thin Martian atmosphere be affecting it? Was the atmosphere actually causing a braking action like the deteriorating orbit of a slowing Earth satellite?

## Phobos Is A Hollow, Empty Tin Can

During an interview about the peculiarities surrounding Phobos, Shklovsky said, "In order to make this braking action so significant, and taking into account the extremely rarefied Martian atmosphere at this altitude, Phobos should have very small mass, that is, very low average density, approximately one thousand times smaller than the density of water."

A density that low, less than an Earth cloud, would have dispersed Phobos eons ago. That could not be the solution.

"But can a continuous solid have such low density, probably smaller than that of air? Of course not! There's only one way in which the requirements of coherence, constancy of shape of Phobos, and its extremely small average density can be reconciled. We must assume that Phobos is a hollow, empty body, resembling an empty tin can."

A tin can? Well, for all intents and purposes the Apollo Lunar Excursion Module was a tin can—exceedingly smaller than Phobos, of course.

## Computer Modeling Of Weird 'Moons'

Shklovsky continued, "Well, can a natural celestial body be hollow? Never! Therefore, Phobos must have an artificial origin and be an artificial Martian satellite. The peculiar properties of Deimos, though less pronounced than those of Phobos, also point toward an artificial origin."

Alien spaceships the size of small moons orbiting Mars? That makes the so-called "Face on Mars" look ridiculously feeble by comparison!

Yet, no less than the United States Naval Observatory weighed in on the Russian astrophysicist's amazing revelation, stating: "Dr. Shklovsky quite correctly calculated that if the acceleration of Phobos is true, the Martian moon must be hollow, since it cannot have the weight of a natural body and behave in the prescribed manner."

Thus, even that august American institution conceded that mysterious alien ships might be orbiting Mars...the objects' strange origins and ultimate purposes completely unknown.

Speculations over what the giant artificial spaceships might be have ranged from massive Martian space observatories, to half-completed generational interstellar spaceships, or even gargantuan planet-killing space bombs left over from an interplanetary war waged millions of years ago.

If they are world-destroying bombs, hopefully they're duds...

## 11. A Famous Astronaut's Last Ghostly Flight

Donald K. "Deke" Slayton was a famous and popular NASA astronaut. One of the original astronauts with the "right stuff," he was part of the Mercury Seven team and a United States Air Force test pilot.

During his last mission for NASA he served as a senior crew member on the historic Apollo-Soyuz joint US-Soviet space docking. Then age 51, he was the oldest man to ever fly in space. (Later that record was broken by Senator John Glenn who returned to Earth orbit aboard the space shuttle.)

Slayton also loved sport planes and he owned a slick racing plane, a fire-engine red Williams Midget Racer, with the registration number N21X.

**Death Of A Hero**

After a lifetime of taking risks many men would shy away from, and some near air crashes, Slayton died quietly in his own Texas home with his wife, Bobbie, and daughter by his side. The official time death occurred was certified as being 3:22 a.m. on June 13, 1993.

And for most men, that would be the end of the story. After all, Deke Slayton had lived a full and amazing life squeezing more things into his 69 years than some others might in five lifetimes.

But Slayton was anything but an ordinary man. He always did what any good test pilot does, he "pushed the envelope." And that's no surprise because test piloting is where that phrase originated.

It seems even death can't separate a pilot and his plane, for just hours after he died, Deke Slayton was seen by dozens of witnesses flying his plane out of an Orange County, California airport.

What happened after his death really didn't shock his family as much as surprise and comfort them.

# MYSTERIES OF THE MULTIVERSE

No, his family wasn't shocked.

The people who were shocked worked for the FAA and the John Wayne Airport in Irvine, California.

## Aviation Enthusiast Martin Caidin Investigates The Mystery

The events that transpired after Slayton's death were first brought to the attention of the world by <u>Martin Caidin</u>, the award-winning best-selling author who penned 50 books, wrote Hollywood adaptations for blockbuster films, and was the creator of the long-running TV series "The Six Million Dollar Man," and its spin-off, "The Bionic Woman."

As Caidin noted in "<u>Of Moon Shots and Ghost Astronauts</u>" Slayton's final flight took place hours after he died.

"Later the same day he died, June 13, 1993 at 7:57 A.M. local time, at John Wayne Airport in Southern California, a Formula One Racing Plane with large FAA-required registration letters and numbers on the fuselage, N21X, took off from the airport and performed various flight maneuvers.

"With a high-speed propeller the extremely noisy aircraft was seen and heard by many people, who clearly identified the type of aircraft and wrote down the N21X registration. The Federal Aviation Administration determined that a noise level mandated by law had been exceeded, and issued a letter of citation against the registered owner and pilot.

On July 20, Bobbie Slayton received a letter in the mail dated June 28, 1993, from the FAA to Donald K. Slayton, notifying him that he was being cited for violating FAA regulations. The letter had been sent to a condo at which the Slaytons sometimes stayed, and its delivery was delayed until Mrs. Slayton picked up the mail there."

## Slayton's Plane Was Placed In Museum Years Before His Death

Responding to the FAA notice, Slayton's widow Bobbie informed them that their allegation was impossible. Not only had her famous husband had died nearly five and a half hours before the supposed violation occurred, but the N21X racing plane was in a museum hundreds of miles from the airport.

Edward Maloney, owner of the Planes of Fame museum in Chino, California got the plane from Slayton who donated it years before his death. Maloney told Caidin: "We've never flown it at all since Deke gave it to us. He was the last one ever to fly it."

## MYSTERIES OF THE MULTIVERSE

Caidin marveled at the eyewitness accounts of people at the airport who clearly saw the plane. Some of the eyewitnesses were the control tower personnel who filed the complaint to the FAA.

"The plane sighted at the airport had taken off on its own. However, to save weight, the plane Slayton flew has no electrical starter. Caidin noted, "and the engine can be started only by a person outside the plane, who swings the propeller while the pilot works the controls inside.

"So how could the plane have been there—hours after Slayton died—while it was in an aircraft museum? And how could it have taken off by itself, with no one outside the plane to start it up for the pilot? If it was Slayton, why did it take so long after he died for the plane to be sighted?

"Bobbie Slayton [humorously] remarked that the reason for the delay Deke took before getting into the air in his racing plane was that 'he probably took six hours to find Gus to prop the plane for him.'" [Gus Grissom, Slayton's best friend, was one of three NASA astronauts tragically killed in the Apollo I fire January 27, 1967]. "

The conclusion of this story is truly remarkable. "Witnesses at the airport who were questioned first by the local authorities, and then by pilots talking to other pilots, and then by still more pilots and investigators sent to Santa Ana for further confirmation, all agreed that the airplane taking off the morning of June 13, 1993, was not only clearly identified as N21X, but that this particular airplane, which had flown for years with this federal registration, was an all-red Formula One racing aircraft, that it departed from the airport in Orange County, flew through various maneuvers in the area, and then flew off in a steady gradual climb on a westerly heading."

The plane flew straight for the Pacific Ocean...gradually left the control tower's radar screen...and was never seen again.

## 12. Did The Nazi's Really Reach The Moon In 1945?

According to a researcher of World War II super weapons, Nazi Germany reached the Moon first. Long before the world was galvanized by Neil Armstrong setting foot on the lunar surface on July 20, 1969, Luftwaffe volunteers orbited the Moon and briefly landed. They didn't walk on the lunar surface as spacesuits hadn't yet been invented.

This incredible tale also claims the disc-shaped space vehicle carried a Japanese officer with orders to report on the achievement directly to the Japanese Emperor, Hiro Hito.

The writer goes on to claim that, "The Germans landed on the Moon in the closing months of World War II, utilizing their top secret, larger exo-atmospheric rocket saucers of the Miethe and Schriever type. The Miethe rocket craft was built in diameters of 15 and 50 meters, and the Schriever Walter turbine powered craft was designed as an interplanetary exploration vehicle utilizing the revolutionary implosion vortex motors invented by engineering genius Viktor Schauberger."

### Nazis On The Moon In 1945?

According to several websites whose writers failed miserably in their due diligence, the Romanian scientist 'Radovan Tomovici', who has studied conspiracy theories for decades, stated, "For over 70 years, it has been common knowledge that the Nazis had a research program overseen by Hans Kammler during the war, with the goal of conquest and control of orbital space. It seems that Kammler, who mysteriously vanished [in Antarctica] shortly before the end of the war, and his team were successful. We're in trouble now."

Unfortunately, the conspiracy theorist bloggers failed to realize that this fascinating tidbit was a piece of PR manufactured for the marketing build up of the film Iron Sky scheduled for an April 2012 release.

## Major Media Gives Nazi Flying Saucers A Second Look

After decades of derision and dismissal by the main stream media, the surviving documents and testimonies by eye-witnesses that a Nazi flying disc program did exist and actually met with some success is being revisited.

The Daily Mail asks: "Hitler's secret flying saucer: Did the Führer plan to attack London and New York in UFOs?" while rival newspaper The Sun claims the Nazi Reich made much more progress with advanced flying saucers than most military historians admitted: "Close encounters of the Third Reich."

Some contend that after the end of the war in May 1945, the Germans continued their space effort from their south polar colony of Neuschwabenland (New Swabia).

Sir Roy Fedden (an aeronautical engineer) stated that the only craft that could approach the capabilities attributed to the flying saucers seen around the world during the late 1940s and early 1950s were those being designed by the Germans towards the end of the war. Sir Roy Feddon , Chief of the Technical Mission to Germany for the Ministry of Aircraft Production, stated in 1945:

"I have seen enough of their designs and production plans to realize that if they (the Germans) had managed to prolong the war some months longer, we would have been confronted with a set of entirely new and deadly developments in air warfare." [Weird World]

One of the strands that feed into ufologists' fertile imaginations is a peculiar structure located outside the village of Nowa Ruba in Poland's Owl Mountains, a part of the country which belonged to Germany up until 1945. Some argue the complex developed Nazi flying saucers.

## A Test Stand For Flying Saucers In Ludwigsdorf, Silesia?

Nick Cook is an aeronautical expert and respected adviser to the worldwide leading journal Jane's Defense weekly. He's also the editor of the air weapons section of Jane's. His book, "The Hunt for Zero Point," dealt with flying saucers, the Nazi disc program and the intense American and British quest for anti-gravity. The remarkable book was an international bestseller.

Cook stated for the record: "It would be a mistake to disregard the research in Germany in the 1930s and 1940s just because it was done in the Third Reich. This kind of suppression of facts would be unscientific and would be just as bad as the suppression of facts that happened during that era."

The Nazi Moon enthusiasts argue that:

- Many of the so-called UFO cases in the 1950s and 1960s—including the famous flyovers above Washington, D.C. during July of 1952—were actually warnings by the Nazi S. S. They flaunted their aerial superiority over the Allies.
- The Nazi S. S. continue to maintain a mammoth, underground city-factory complex far beneath the frozen wastelands of the Antarctic ice not far from U. S. and Norwegian bases.
- The disciplined remnants of an elite Waffen S. S. corps and their appointed successors monitor the superpowers and continue an ingenious yet relentless campaign (begun in early 1947) of military, economic and political blackmail and extortion on a worldwide scale.
- The diabolical Nazi S. S. established and expanded upon a network of lunar bases whose primary purpose is to exploit the ultimate military "high ground," control and contain the superpowers, and mine the abundant minerals available there (including Helium-3 for nuclear fusion reactors) for light and heavy manufacturing purposes to support terrestrial and lunar operations. Some contend this is why NASA killed the Apollo program and never returned to the moon. Many space probes have returned photos of what can only be described as mining operations and structures on the lunar surface. Are they operated by space aliens? Not likely. More likely the Nazi space corps run the show.
- The Nazi S. S. established secret maintenance bases in extreme northern Norway, under deep lakes located within northwest Canada, Central America, southern South America, and under targeted regions of the vast oceans and seas. These areas have been hotbeds of flying saucer reports and UFO activity for years, but strangely no activity was ever reported from these regions prior to 1945.
- When Admiral Richard E. Byrd's United States naval task force confronted and engaged the Polar Waffen SS forces near the giant Nazi base of Neuschwabenland in the Antarctic in 1956 the Americans lost. The task force limped back with home months earlier than planned with almost one hundred men missing and scores more injured. One of the fleet's ships was lost and Byrd was ordered by President Eisenhower never to talk to the news

media about it the horrifying events that transpired on the southern ice pack.

• The German navy's "super submarines" disappeared at the fall of the Third Reich. They were never sunk or scuttled and none of the Allied forces captured them. They were last reported on a heading towards the South Pole.

• Tons of Nazi gold disappeared at the end of the war. The Allies never found it, nor did anyone else. Today the horde has an estimated value of trillions of dollars. Did the so-called Nazi "Last Battalion" smuggle it out of Germany before the Allied Forces arrived? Did they use it to further capitalize their Antarctic base and satellite operations in Argentina and the mountainous fortresses they built that still are in operation today in Paraguay?

Are these theorists correct? The answer may lie on the far side of the Moon.

MYSTERIES OF THE MULTIVERSE

# STRANGER THAN FICTION

## 13. Doomsday: Earth's Core Spinning Out Of Control?

No one wants to think about it, but strange core mutations may signal an expanding Earth Doomsday.

Nearly 700,000 years ago the Earth's core shifted, changed the magnetic field, and precipitated mass extinctions around the globe. Now worried scientists think it may be happening again.

**The Expanding Earth Theory, And Core Mutations: An Inevitable Doomsday?**

The theory of the expanding Earth has fallen in and out of favor over the years. Currently, most geophysicists discount the idea. They point to a lack of evidence supporting such a thing.

Yet, advocates counter by pointing to deep ocean fissures that continue to split and widen while being "patched" by material upsurging from the planet's mantle. It's evidence, they argue, that the Earth is expanding, blowing up like a balloon, and will eventually explode many millions of years from now.

The shifting, aberrant, unstable core is a symptom of the process and causes geomagnetic field instability.

It's another step along the road towards Doomsday.

According to expanding Earth theorists, the planet will grow bigger and bigger until it bursts.

That day the Earth will cease to exist, continuing in orbit around the sun merely as tumbling chunks of jagged rock and clouds of drifting dust.

**Unstable Core A Symptom Of Inflation**

## MYSTERIES OF THE MULTIVERSE

Theorists that support the expanding Earth scenario promote the idea that a planet once existed between Mars and Jupiter—a region where only shards of rock remain. Known now as the asteroid belt, they claim that once a giant planet orbited there and expanded until it blew up.

Mars was a moon of that planet and other chunks of what remained either fell into the sun, were catapulted into space, or were captured by the large gas planets becoming additional moons.

Only a fraction of the planet remained in its original orbit. The asteroids are the debris of—and a testament to—the death of a world.

Believers of the expanding Earth hypothesis argue that none of the large dinosaurs could have existed on today's world. Earth's gravity is too high now for the larger dinosaurs to have survived. Their weight and mass was adapted to a planet one-third its current size with a gravity field less than half of today's.

The continents drifted away from each other not by continental drift of the underlying tectonic plates, but because as the Earth inflates the land masses naturally are pushed farther apart.

Paleontologists and geologists scoff at the theory and point to evidence they have amassed discrediting such things.

### Core The Key To Inflation

But the advocates of the expansion theory are persistent. The core and its permutations drive the expanding Earth, they say. And now as the core wobbles and shifts—causing the entire geomagnetic pole to shift—they have redoubled their arguments and their critics have no response.

The core shift affects the deep sea plates—the mega-continent sized plates the land masses sit on. As the core shifts the mantle pressure builds and the plates crack and move generating mega-thrust quakes like the ones that devastated Japan during March 2011.

And as the magnetic field shifts it causes an electromagnetic flux affecting climate.

Its interaction with the sun's electrical field, they claim, will fuel the outbreak of superstorms across the planet from Russia to Australia, from the United States to northern Europe.

And there is evidence of that with the two back-to-back superstorms that pummeled Australia, the giant storms that lashed the UK and Russia and the titanic snowstorms that slammed America during the winter of 2010.

Now during early 2011 more superstorms have whipped across Europe and ferocious storms have pounded the United States. During April 2011 alone more than 600 tornadoes devastated huge regions of Texas and the South up towards the mid-Atlantic states. At one point the storms affected many millions in more than 25 states and caused terrible destruction across 14 of them. Hundreds of people lost their lives and property damage soared into the billions.

Some scientists see this as only the beginning and predict bigger and more frequent storms.

And all the while the core continues its turbulent shifting. Writhing and spinning out of sync with the rest of the planet...a sphere of molten metal gone mad as the Earth expands...

MYSTERIES OF THE MULTIVERSE

# 14. The Mysterious Prehistoric Ropen Of Papua New Guinea

The Ropen ("demon flyer") is a creature that has terrorized the natives of Papua New Guinea's jungles and outlying islands for thousands of years.

First brought to the attention of Western missionaries following the second world war, the nocturnal flying creature is said to have large bat-like wings, an elongated beak filled with teeth, sharp, tearing claws and a very long tail with a split or flange on the end. Reports from natives and a few investigators claim that the creatures also glow in the dark and are visible at night in the sky. This phenomenon, called the "Ropen light," was observed and videotaped by researcher David Woetzel.

Researchers speculate that the bio-luminescent glow aids the creatures in catching fish, their primary diet.

### The Ropen And The Duah

Eyewitness reports collected by exploration teams seeking evidence of the creature leads researchers to believe there may be two distinct animals.

This hypothesis is supported by a 1595 maritime chart cautioning sailors about sea monsters and depicting coastal regions of the Earth where they might be found. In the area of Papua New Guinea, two 'sea monsters' are illustrated, one is much larger than the other, yet both have similar physical traits. Each have long necks, prominent head crests, tails ending with a flipper and ridges along their backs. They are shown flying above an island.

Debates usually accompany any cryptozoological animal and the Ropen is no different. The general consensus amongst orthodox scientists is the creature doesn't exist. Amongst those that have actually traveled deep into Papua New Guinea's rain forests and tiny offshore islands, not only do two creatures exist, but strong evidence exists that one is very large with a 20-foot or larger wingspan while the other has only a four to four and a half foot wingspan.

The natives agree. They call the small creature a "Duah," and claim it's not a Ropen at all.

### Can The Creatures Be Pterosaurs?

Researchers believe both are members of the pterosaur family thought to be extinct some 65 million years ago.

The large Ropen may be living Dimorphodon pterosaurs, with dermal bumps and a head crest, while the descriptions of the smaller creatures that presumably inhabit the caves of the islands dotting the Bismarck Archipelago tally with that of the extinct Rhamphorhynchus, a pterosaur with a wingspan of three to four feet.

A creature spotted flying around Manus Island off Papua New Guinea with a wingspan of three to four feet, according to Jim Blume, a missionary in Wau, on the mainland may be a Rhamphorhynchus.

Blume's investigations indicate that wingspans may reach ten to fifteen feet in other areas. Several other investigators mention native encounters with Ropens that have much larger wingspans.

Whatever the two creatures are, they may not be exclusive to Papua New Guinea, similar creatures have been sighted and reported for centuries in Central Africa. The natives in the Congo, Zambia, Angola, Kenya and Zaire call it the "Kongomato." They claim that the reddish creature has large, leathery wings, sharp claws, a split tail, and teeth.

## 15. The Ahool: The Monster Mega-Bat That Roams The Skies Of Java

"The jungle has a thousand ways to kill you and none of them are pleasant." ~ Anonymous

The gigantic Bismark flying fox is officially the largest bat known to exist. It makes its home in Asia.

Yet a true monster bat may exist—a virtual flying terror—the mega-bat called the Ahool. This nightmarish creature is sought by some of the world's top cryptozoologists.

With a body the size of a small child, and a reported 12- to 15-foot wingspan, the Ahool dwarfs all other bats. It's said to viciously attack livestock and men and has been reported to feed off human infants, although by most accounts it primarily eats fish.

### First Westerner Witness

Although seen and reported for centuries by natives inhabiting the region of Java where the creature is said to dwell, the first Westerner witnessed the Ahool during the early 20th Century.

The year was 1925 and Dr. Ernest Bartels—son of the noted ornithologist M.E.G. Bartels—recorded the first known description of this legendary Javanese beast. According to the doctor's notes, he saw a gigantic bat-like creature swoop over a waterfall in the foothills of the Salek mountains.

Several years after the initial encounter, Bartels again spotted the monster. That time he also heard its distinctive cry: "a-hooool." The creature's name was adopted from its hunting cry.

## Physiology Of The Creature

As described by natives, the Ahool appears like a monstrous bat-like animal about the size of a one-year old human. Although all agree the creature has the body of a bat, some insist that its head is more similar to that of a monkey's than a bat's. A few descriptions from eyewitness testimony over the years even describe the face of the creature as appearing eerily human.

The face is also said to have two large, shiny black eyes.

Its wingspan has been estimated between 12- to 15-feet, although most reports peg it as no more than 12-feet. Two flattish forearms support the immense wings.

Curiously, many of the documented sightings report that the creature's feet appear to be pointing backwards.

A thin, darkish coat of hair covers the creature. Most reports describe the color as gray, although some witnesses swear it's brownish. The discrepancy in color may be due to the angle of the sun or shadows during the sightings.

## The Ahool's Habits And Habitats

The monster bats are mostly nocturnal. Sightings during the daylight hours are exceedingly rare.

According to natives familiar with the creature, the Ahool nests deep inside natural caverns and is sometimes found in hollows behind waterfalls. It prefers the area along rivers. At night it hunts for prey above the rivers.

Cryptozoologists speculate that the Java animal may be similar to the Kongamato of Africa. The Kongamato—another species of gigantic jungle bat—is reportedly smaller, has red-hued fur and a protruding, pointed snout.

Famous researcher Ivan T. Sanderson investigated the reports of the Ahool extensively and came to the conclusion that the creature was a previously unknown gigantic bat.

# 16. Man-Eating Giants Discovered In Nevada Cave

Many Native American tribes from the Northeast and Southwest still relate the legends of the red-haired giants and how their ancestors fought terrible, protracted wars against the giants when they first encountered them in North America almost 15,000 years ago.

Others, like the Aztecs and Mayans recorded their encounters with a race of giants to the north when they ventured out on exploratory expeditions.

Who were these red-haired giants that history books have ignored? Their burial sites and remains have been discovered on almost every continent. In the United States they have been unearthed in Virginia and New York state, Michigan, Illinois and Tennessee, Arizona and Nevada.

And it's the state of Nevada where the story of the native Paiute's wars against the giant red-haired men transformed from a local myth to a scientific reality when in 1924 when the Lovelock Caves were excavated.

At one time the Lovelock Cave was known as Horseshoe Cave because of its U-shaped interior. The cavern , located about 20 miles south of modern day Lovelock, Nevada, is approximately 40-feet deep and 60-feet wide.

It's a very old cave that pre-dates humans on this continent. In prehistoric times it lay underneath a giant inland lake called Lahontan that covered much of western Nevada. Geologists have determined the cavern was formed by the lake's currents and wave action.

**The Legend**

The Paiutes, a Native-American tribe indigenous to parts of Nevada, Utah and Arizona, told early white settlers about their ancestors' battles with a ferocious race of white, red-haired giants. According to the Paiutes, the giants were already living in the area.

The Paiutes named the giants "Si-Te-Cah" that literally means "tule-eaters." The tule is a fibrous water plant the giants wove into rafts to escape the Paiutes ' continuous attacks. They used the rafts to navigate across what remained of Lake Lahontan.

According to the Paiutes, the red-haired giants stood as tall as 12 feet and were a vicious, unapproachable people that killed and ate captured Paiutes as food.

The Paiutes told the early settlers that after many years of warfare, all the tribes in the area finally joined together to rid themselves of the giants.

One day as they chased down the few remaining red-haired enemies, the fleeing giants took refuge in a cave. The tribal warriors demanded their enemy come out and fight, but the giants steadfastly refused to leave their sanctuary.

Frustrated at not defeating their enemy with honor, the tribal chiefs had warriors fill the entrance to the cavern with brush and then set it on fire in a bid to force the giants out of the cave. The few that did emerge were instantly slain with volleys of arrows. The giants that remained inside the cavern were asphyxiated.

Later, an earthquake rocked the region and the cave entrance collapsed leaving only enough room for bats to enter it and make it their home.

**The Excavation**

Thousands of years later the cave was rediscovered and found to be loaded with bat guano almost six feet deep. Decaying bat guano becomes saltpeter, the chief ingredient of gunpowder, and was very valuable.

Therefore, in 1911 a company was created specifically to mine the guano. As the mining operation progressed, skeletons and fossils were found.

The guano was mined for almost 13 years before archaeologists were notified about the findings. Unfortunately, by then many of the artifacts had been accidentally destroyed or simply discarded.
Nevertheless, what the scientific researchers did recover was staggering: over 10,000 artifacts were unearthed including the mummified remains of two red-haired giants—one, a female six and a half feet tall, the other male, over eight feet tall.

Many of the artifacts (but not the giants) can be viewed at the small natural history museum located in Winnemucca, Nevada.

### Confirmation Of The Myth

As the excavation of the cave progressed, the archaeologists came to the inescapable conclusion that the Paiutes myth was no myth; it was true. What led them to this realization was the discovery of many broken arrows that had been shot into the cave and a dark layer of burned material under sections of the overlaying guano.

Among the thousands of artifacts recovered from this site of an unknown people is what some scientists are convinced is a calendar: a doughnut-shaped stone with exactly 365 notches carved along its outside rim and 52 corresponding notches along the inside.

But that was not to be the final chapter of red-haired giants in Nevada. In February and June of 1931, two very large skeletons were found in the Humboldt dry lake bed near Lovelock, Nevada.

One of the skeletons measured eight and a half feet tall and was later described as having been wrapped in a gum-covered fabric similar to Egyptian mummies. The other was nearly ten feet long. [Nevada Review-Miner newspaper, June 19, 1931.]

## 17. How Nikola Tesla Almost Destroyed The World

During 1908 the quiet Siberian countryside was rocked by a blast more powerful than the atomic detonation set off over Hiroshima, Japan.

In literally seconds, 800 square miles of virgin forest lay splintered and flattened in the Tunguska region of Russia. The horrific blast eventually became known among investigators as the "Tunguska Event."

What caused such violent and widespread devastation? The most widely accepted theory amongst orthodox scientists is that a cometary or meteoric explosion occurred in the upper atmosphere. The resulting shock waves flattened everything below the explosion.

Yet, that explanation may be misleading. Other evidence exists that supports a different reason for the event.

And that evidence leads to a markedly different cause—a cause that is stunning: world famous inventor Nikola Tesla accidentally set off the explosion while testing a powerful energy broadcasting device. Later, Tesla would make oblique references to the technology he created as a 'death ray' and urged its use as a military weapon.

If the Tunguska destruction was man-made it would seem to mesh with much of the eye-witness testimony gathered in the aftermath of the event.

This alternative theory has been promoted during the past several years by Oliver Nichelson and others.

### Testimony of Russian Eye-Witnesses

Accounts gathered by the Russian mineralogist Leonid Kulik, in his 1930 expedition to the site of the explosion are consistent enough on many details to be considered generally reliable. Most recalled a bluish-white cylinder of light in the sky followed by a series of concussive reports like thunder. A few reported the ground trembling as if dozens of freight trains ran underneath the ground. The significance of that is they felt the rumbling vibration coming from beneath the ground before the explosion.

These reports could all support a multibillion watt energy pulse emerging from the ground as a bluish-white light of such intensity it outshone the sun. It was cylindrical in appearance and caused a series of thunderous reports such as lightning strikes create during violent electrical storms.

The Russian newspaper Krasnoyaretz reported on July 13, 1908 just two weeks after the event: Kezhemskoe village: An unusual atmospheric event was observed. At 7:43 a.m. the noise akin to a strong wind was heard. Immediately afterward a horrific thump sounded followed by an earthquake that literally shook the buildings as if they were hit by a large log or a heavy rock. The first thump was followed by a second and then a third. Then the interval between the first and the third thumps were accompanied by an unusual underground rattle, similar to a railway upon which dozens of trains are traveling at the same time. Afterward for 5 to 6 minutes an exact likeness of artillery fire was heard: 50 to 60 salvos in short, equal intervals, which got progressively weaker. After 1.5 - 2 minutes after one of the "barrages" six more thumps were heard, like cannon firing, but individual, loud and accompanied by tremors.

The sky, at the first sight, appeared to be clear. There was no wind and no clouds. However upon closer inspection to the north, i.e. where most of the thumps were heard, a kind of an ashen cloud was seen near the horizon which kept getting smaller and more transparent and possibly by around 2 - 3 p.m. completely disappeared.

Other first-hand accounts from eyewitnesses corroborate the evidence investigators discovered when sifting through the debris of the event's aftermath. Many described a darkened cloud that gradually flattened into a dish or saucer-shaped form. The cloud was pierced by an intense beam or shaft of light.

Testimony of witnesses: Kirensk, a farmer - "[I saw] a fiery pillar in the form of a spear"; Nizhne Karelinsk, a chicken breeder—"[It] turned into a fiery pillar and disappeared in a moment"; Another unnamed farmhand—"A forked tongue of flame broke through the cloud"; from a man identified only as Vanavera interviewed by the Krasnoyarsk newspaper—"A huge flame shot up and cut the sky in two."

There's something else that's odd that does not lend itself to supporting the popular theories of meteoric, asteroid or cometary airborne explosion. The day of the Tunguska Event the Irkutsk Observatory recorded magnetic anomalies that left a signature resembling those that are made by atomic blasts. The detection of the magnetic disturbances began about six minutes after the initial explosion over Siberia and continued for more than four hours. The blast signatures that the observatory recorded are almost identical to those recorded decades later that nuclear air bursts generate.

Many scientists have argued that if the Tunguska Event was caused by a meteor, asteroid or comet those signatures could not have been produced. It is recognized that some large meteors passing through the Earth's ionosphere have caused minor magnetic disruptions, but none compares to the four hour plus magnetic disturbance generated by the Tunguska explosion.

**Tesla And Death Rays**

Nikola Tesla (1856 - 1943) is perhaps the greatest overlooked genius in American history. His inventions are legion and his investigations into the nature of electricity and magnetism are still finding applications today.

Among his many accomplishments, Tesla developed the technology that enabled television to become a reality; he enabled Edison's power plants to transmit electricity 1000 times farther than Edison's method; and he built and tested radio long before Marconi. In one of those flukes of history, Tesla—a perfectionist—finally brought his radio transceiver (far superior to Marconi's rudimentary device) to the U.S. Patent Office two days after Marconi's application. The patent, of course, was awarded to Marconi.

Among Tesla's many inventions was broadcast power. His devices enabled machinery to run without being plugged in to an electrical grid. In his world, the entire Earth was an electrical grid.

While his broadcast power experiments made world news, his greatest project—one that later led to his infamous death ray experiments—was the broadcast tower in Colorado Springs. That tower was the precursor to his Wardenclyffe Tower project in Shoreham, Long Island, New York that was never fully completed.

In a letter to the New York Times dated April 1908 Tesla expanded upon his idea of destruction by electrical beams. He wrote, "When I spoke of future warfare I meant that it should be conducted by direct application of electrical waves without the use of aerial engines or other implements of destruction." Then he went on to add, "This is not a dream. Even now wireless power plants could be constructed by which any region of the globe might be rendered uninhabitable without subjecting the population of other parts to serious danger or inconvenience."

Tesla knew what he was talking about. He had constructed such towers and seen first-hand what they could do. They were capable of generating great destructive power arriving at the speed of light anywhere on the Earth.

Several theorists have proposed that Tesla was testing his wireless power generator during June of 1908. They point to some interesting facts to support their case. According to one, "Historical facts point to the possibility that this event was caused by a test firing of Tesla's energy weapon."

He draws this conclusion based on the fact that Tesla wrote at length about the powerfully destructive ability of his new energy transmitter. Based on the designs of the prototype that Tesla had built and tested in Colorado Springs, his Wardenclyffe complex and primary energy tower were a quantum leap beyond his original transmitter in Colorado.

Tesla's tireless effort to improve upon his generation of wireless energy continued to mount during 1900 to 1910. According to some historical students of Tesla's life they contend the brilliant inventor had reached a desperate crossroads in his life: facing mounting financial woes and at loggerheads with orthodox scientists, Tesla fell into a deep depression and suffered a nervous breakdown. Some theorize that in a dramatic bid to resurrect his formerly brilliant career, the inventor might have tested his giant transmitter to demonstrate its massive destructive ability.

The year most likely for that demonstration to have occurred? 1908.

Years later, during 1915 Tesla wrote: "It is perfectly practical to transmit electrical energy without wires and produce destructive effects at a distance. I have already constructed a wireless transmitter which makes this possible. [But] when unavoidable [it] may be used to destroy property and life. The art is already so far developed that the great destructive effects can be produced at any point on the globe, defined beforehand with great accuracy."

When Tesla wrote that in 1915 he seems to have made an admission of a test of the tower. Although undocumented, Tesla did have the capability to transmit high energy wave frequencies generating catastrophic forces in excess of 10 megatons.

Directed wireless power transmitted through the globe and erupting with raw fury from the surface somewhere else is perfectly consistent with the evidence of the aftermath of the Tunguska Event and—more importantly—it was supported by eyewitness accounts of the catastrophe.

Lingering Clues after the Event

No professional or amateur astronomers anywhere on Earth reported a fireball. A gigantic booming sound was heard with multiple reports diminishing in intensity afterward; no debris fell from the sky. No impact crater has ever been found which is reasonable if an energy beam erupted upwards from beneath the ground. Yet, magnetic and electrical disturbances were reported for several days over Europe. The sky glowed like twilight all night long. Massive glowing "silvery clouds" were reported over northern and northeastern Russia.

All of this is consistent with significant electrical disturbances of the atmosphere which Tesla's massive Wardenclyffe Tower was quite capable of achieving.

Tesla claimed that his wireless transmitter achieved experimental power levels into the tens of billions of watts. That power, released within a time frame compressed to microseconds could easily have achieved the destructive energy of a multi-megaton explosion. As some have speculated, what Tesla had created was a diabolical device capable of transmitting the destructive power of hydrogen bombs. As he later suggested through his letters and newspaper interviews any location anywhere on the planet could be annihilated at the speed of light and with the mere flick of a switch.

**Did Tesla Destroy The Tunguska Forest?**

Could any of this actually be true? Was Tesla exaggerating the power of the tower or was he under the veil of a self-delusional fantasy?

Recently, a team of electrical engineers conducted a wide-ranging analysis of Tesla's wireless transmission technology. His machine did not propagate radio waves as we understand them today. Instead, the tower—and the prototype in Colorado Springs—transmitted electrostatic energy waves. Those waves could easily pass through the Earth. Little power would be lost during the pulse transmissions.

Did he power up and "fire off" a transmission towards Russia? According to some researchers, circumstantial evidence discovered amongst Tesla's notes, the chronology of his work and his financial upheavals point to the strong possibility that he did fire the wireless energy from the tower at full power at least once and maybe on several other occasions at lower power levels.

The evidence, while not incontrovertible, does point towards the possibility that Tesla tested his wireless energy generator. If he did, the test would have occurred sometime around the middle of 1908. A likely target could have been the sparsely populated Polar Regions. Tunguska sits on the southernmost edge of the Arctic Circle.

So what caused the Tunguska Event—a meteor, asteroid, comet, methane gas explosion...or Nikola Tesla's Wardenclyffe Tower project? Each hypothesis has evidence for and against it. Did Nikola Tesla really shake the world more than 100 years ago?

That question remains unanswered.

## 18. Thomas Edison's Conversations With Ghosts

During his lifetime, Thomas Alva Edison filed for almost 1,700 patents. 1,093 applications were successfully awarded the prodigious inventor.

Yet in the archives of his achievements at the Edison National Historic Site, housing five million pages of documents, none of them mention—and no where can be found—a patent, schematic or even a scribble for a paranormal invention that is still attributed to Edison today: the psycho-phone.

### A Spirit Phone?

The psycho-phone is a curious device with an even curiouser history. Said to be the culmination of the inventor's research into the other world, or life after death, the psycho-phone allegedly gave the user access to the dead. Supposedly loved ones could be contacted through it or the brains of famous men such as Aristotle or Julius Caesar could be picked—assuming the user was able to converse in ancient Greek or Latin.

The story of Edison's quest to invent a machine that could place phone calls to the dead is a murky one. It can be traced, however, to an interview he gave to "Scientific American" during the 1920s.

During the conversation with reporter B.F. Forbes from the magazine, Edison confided he was working on a machine that could contact the dead. The story made headlines in many papers across America.

### Excerpt From "Scientific American" Interview

"If our personality survives, then it is strictly logical or scientific to assume it retains memory, intellect, other faculties and knowledge we acquire on

Earth. Therefore...if we can evolve an instrument so delicate as to be affected by our personality as it survives in the next life, such an instrument, when made available, ought to record something."

This intrigued the reporter who asked Edison to expound upon the idea. Edison obliged him saying "it might be possible to construct an apparatus which will be so delicate that if there are personalities in another existence or sphere who wish to get in touch with us in this existence or sphere, this apparatus will at least give them a better opportunity to express themselves than the tilting tables and raps and Ouija boards and mediums and the other crude methods now purported to be the only means of communication."

Reading "Edison Working to Communicate with the Next World," it's easy to see the reporter attempted to confirm that Edison was working on technology designed to make contact with the beyond and slanted the whole article towards that end.

Edison said, "I don't claim that our personalities pass onto another existence. I don't claim anything, because I don't know anything ... for that matter, no human being knows. But I do claim that it is possible to construct an apparatus which will be so delicate that if there are personalities in another existence who wish to get in touch with us ... this apparatus will at least give them a better opportunity."

Excerpts from the interview made the front pages of newspapers around the globe.

Several years later, Edison admitted to the press that he'd been making a joke at the Scientific American reporter's expense. He asserted he had never worked on such a device.

### The Myth

A story emerged claiming that after the interview, Edison began a dialog with Sir William Crooke. The Englishman had invented the vacuum tube that Edison later used to create the incandescent light bulb.

Crooke was a spiritualist that was deeply immersed in the paranormal—especially what was then called "spirit photography."

Supposedly Edison was influenced by this and commented that if spirits could be captured on film, then it might be possible to communicate with them.

While Edison was a well-known agnostic, he was also a spiritualist of sorts. During interviews and speeches, he sometimes shared his personal views of an afterlife. He envisioned a limbo-like dimension where the core of every person's intelligence and personality waited before moving on to another phase.

The story continues as Edison, attacking the problem in his usual workaholic way, focused intently on creating a working model of what he dubbed the "psycho-phone."

Joseph Dunninger, a famous magician of the time, testified that Edison had demonstrated a working prototype of the apparatus. Perhaps conveniently, he did not make this claim until after the inventor's death.

Soon after Edison's passing a concerted search was made for the psycho-phone. None could find the device, a model or even a rough draft. Nothing relating to the machine has ever been found.

But the story of the mysterious psycho-phone did not end with the futile search. A decade after the inventor's death, Edison was said to have made contact with a medium during a seance. The medium asserted the plans could be found with three of Edison's assistants. Dutifully, the assistants were found and the machine reportedly built.

It didn't work.

Edison (presumably a bit flustered by this time) made another appearance at another seance. This time the inventor imparted tips on how to get the device working. J. Gilbert Wright, the inventor of putty, happened to be at the seance. He scribbled notes and began working on the psycho-phone. After a few more engineering consultations with various spirits, Wright continued to tweak the machine hoping to make contact with...somebody. His work continued without success right up to his death in 1959.

Then the infamous psycho-phone vanished.

When considering the time lines, the players, the events and the nature of the quest, the entire affair rings of a long, drawn out hoax. No proof is forthcoming that the psycho-phone ever existed.

Perhaps some day a researcher will actually invent such a device. If that happens, maybe someone will think of calling up Edison and inquiring about the infamous psycho-phone.

Imagine what the long-distance bill on that call might be...

## 19. The Chilling Mystery Of The Five Indian Maidens

In the annals of North America few tragedies equal the poignancy and tragedy of the five Indian maidens. Fewer yet had a chilling unsolved mystery attached to them.

The tragedy that haunts the memories of the people living outside Dog Creek, British Columbia was discovered by a Major Fredrick Brewster during the early 1950's. Brewster was a backwoods guide and also outfitted hunters with supplies needed for treks into the Canadian wilderness.

During a trip into the mountains above the town of Dog Creek, Brewster found five graves under an old evergreen tree. Graves in that area of Canada were rare, especially at altitude in the mountains, so his curiosity was piqued. He wondered if they were the graves of settlers attacked by an Indian raiding party, or perhaps a small Indian grave site.

Some years passed while Brewster grew his business and sporadically gathered what information he could about the five unmarked graves high above the remote town of Dog Creek.

Nearly ten years went by before Brewster cobbled together the truth about who occupied those graves and the underlying tragedy that led to their presence on that lonely, windswept mountainside.

His investigations revealed a French-Canadian trapper from Quebec who arrived in British Columbia during the 1880s with a wave of other Europeans. What drew those rough and tumble entrepreneurs and adventurers were thoughts of quick fortunes for the asking—primarily from gold or furs.

The Frenchman quickly became known in the area as an amiable, earnest young man.

Once he'd settled in with others of his ilk he began to become friendly with an Indian tribe not far from the town where he lived. Soon he was accepted and welcomed as a brother.

It so happened that this particular tribe, an offshoot of the Assiniboine Indians to the east, had many young, single women but very few young men. Many of the younger men had been killed in tribal warfare or raids against white settlements before peace had broken out—a peace enforced by the Northwest Mounted Police.

Because of this dearth of available young braves, the comely Indian maidens began to develop an interest in the handsome young white man who regularly paid visits to their village. The trapper returned their interest in full measure. Thus, in a fairly short period of time, five of the young women had fallen in love with him.

Unfortunately, the man was a scoundrel. Encouraging their feelings, he then took full advantage of them and slept with all of them as often as he could. Yet he kept from them the knowledge of the other women, so that each of his lovers thought she was the only one receiving his favors. How he kept up this precarious balancing act is unknown. Brewster was unable to discover that.

Eventually several of the girls became pregnant exposing the Frenchman's charade.

Becoming suspicious of the intricate relationships he had developed and the twists and turns he had mastered to keep them all at arm's length from each other, the five finally sat down together to share stories about their lover.

Imagine their dismay and dishonor when they discovered the fraud. They'd been tricked and used. Worse, they had been despoiled in the eyes of their tribe and by custom. Their shameful actions could never be erased. Their children would become outcasts.

The young girls could not face the Frenchman, even to confront him with the evil he'd done. Nor could they face their people and endure the shame and scorn their actions had earned.

According to Brewster, that very day they made a death pact between themselves and, packing provisions, set off together for the mountains. He adds that although the young women were no longer virgins they would have still been considered maidens in the eyes of their tribal elders.

When they reached a lonely, tranquil spot in the mountains they stopped. That was journey's end; the place they chose to embrace Death as their new lover. That idyllic mountainside was their avenue of escape from the vilification they would have endured had they stayed with their tribe.

They shared only one rope between them: one rope for five thin, lovely necks; one rope to end the lives of five teenagers whose grief and terror and hurt and shame were far too great to live with.

Death was a greater comfort.

As fate would have it, a Northwest Mounted Policeman stumbled across the maidens' bodies next to the hanging tree; the solitary rope was still in place.

Each of their young bodies had been laid side-by-side as if they had settled down to rest together. Reconstructing what had occurred, it was obvious that as each had hanged herself in turn, one or more of the others lowered the body to the ground and laid it out next to the tree.

But the nagging question remained—the unsolved riddle. Who had taken down the body of the last girl and stretched her out next to the others? All had abrasive rope burns around their necks. All had died by hanging.

Try as they might, the Mounties found no evidence anyone else had been within miles of the mass suicide. No telltale hoof marks or any other evidence was found that some mysterious fifth person had been there and gone. Yet someone had to have lowered the fifth girl to the ground.

To this day no one has been able to provide an answer to the enigma.

## 20. Tragedy Of The Wailing Watcher Of The Waves

Many unexplained mysteries haunt the nooks and crannies of the islands of the Caribbean. Tales of pirates' deeds and misdeeds abound, as well as those unnerving stories best told late at night in whispers over large measures of strong rum.

The isle of Barbados is no exception. Some say the ghost of Sam Lord has been seen walking the hallways of his stately mansion. Others still relate the famous story of the Chase family's burial vault and the coffins that seemed have have a will of their own.

Yet perhaps none of those stories are as intriguing, nor as poignant, as that of the sailor Cornelius Barnesmythe.

Barnesmythe sailed the Caribbean for more than thirty years on frigates and merchant vessels. For a brief time he signed aboard a whaling vessel but found the cold weather and rougher seas didn't suit him.

The last voyage he ever made was in 1879. His ship had just passed the tiny island of St. Kitts when he spied something in the moonlight on the shore.

What he saw eventually drove him mad, for Cornelius Barnesmythe had spotted a mermaid. If that had been all there was to it the poor man no doubt would have kept his faculties, but as he later admitted during jabbering testimony at a Maritime hearing held in Port of Spain, Trinidad, when he saw the mermaid—the most beautiful creature he had ever cast his eyes upon—he became instantly entranced and a compulsion came over him to jump overboard to reach her.

And that he did.

The captain and crew of his ship didn't notice him missing for several hours as Barnesmythe was off-watch and no one monitored his activities.

When they did discover they had a man overboard they turned the ship about and launched a search for their shipmate.

Dawn broke. All that morning and well into the afternoon they searched. Although the currents could have taken Barnesmythe far outside the shipping lane they stubbornly pressed ahead.

By dusk they'd all but given up hope when a seaman cried out. His sharp eyes had spotted movement along a thin strip of sand between the rocks along the southern shore of St. Kitts. A skiff was launched and Barnesmythe was rescued.

Barnesmythe was a man possessed. He related a strange tale of finding a mermaid on the island who had called him to her. He swore she had stolen his soul. He begged to be returned to the island. He could not bear to be departed from her.

The captain ordered the crew to detain Barnesmythe until they could reach port. When they made landfall the miserable sailor was dragged away sobbing and wailing.

The hearing found him guilty of abandoning his ship and endangering the crew. It also determined that he'd lost his ability to reason and therefore his seaman's certificate was voided. The court decided against jailing him as they considered him mentally unstable and not responsible for his actions.

Somehow, Barnesmythe made his way from Port of Spain, Trinidad to Bridgetown, Barbados. From there the records show he built a one-room hut of coral and cement perched on top of a giant rock where the land met the sea.

Barnesmythe lived there until the end keeping a twenty-four hour vigil: watching the ocean; watching the horizon, waiting for his elusive the mermaid to return to him.

For many years the locals who passed that way at night heard ululating moans drifting from the hut—the mad sailor bewailing his lost mermaid who never returned.

MYSTERIES OF THE MULTIVERSE

## 21. The Howling Horror In Harrison's Cave

Harrison's Cave is one of Barbados's biggest tourist attractions. Ask any of the guides that work there about the creature that inhabits its Byzantine depths and they'll dismiss the story as sheer nonsense and laugh.

But it was no laughing matter to Johnny Beldfourd, an island worker contracted to do some masonry work deep inside the cavern. During the early 1960s the government hired his company to shore up a section of the caverns not open to the public.

As he was working one morning he heard a scuffling noise. Looking up he saw a dark figure crouching against one of the cavern walls. He swung his work light about and pinned the figure in its beam.

What he saw eventually drove him mad. In his more lucid moments before being institutionalized at a mental health facility, he described seeing a horrid creature "the likes of which no God-fearing man should ever see."

According to Beldfourd, the creature—some sort of cave dwelling troglodyte—stood about four feet tall. Beldfourd was uncertain about the height because for the thirty seconds he encountered the creature it remained in a crouched position. It carried the carcass of a small animal—perhaps a cave bat—and its slit of a mouth and yellowed fangs were smeared with blood.

Both the creature and Beldfourd seemed equally shocked by their mutually unexpected encounter.

Later, the shaken man claimed the thing stank and it made deep guttural noises. The only other characteristics he could recall were the creature's eyes—which were shaped like huge saucers—and testified its hairless body had "the color of curdled cream."

Although the cavern was thoroughly searched by the authorities nothing was found. Island experts tried to place the animal that Beldfourd insisted he saw that morning. Nothing came close in the animal world, let alone anything indigenous to the island.

As most incidents of this nature—especially one without any corroborative witnesses—the whole incident was written off as a hoax, or the ravings of a lunatic.

The latter must have been the consensus of the authorities because within a few weeks the hapless Beldfourd was ordered by a court to be admitted to a hospital. Beldfourd was sent and there he remained until he died.

Terrifying cave creatures have been reported throughout history. What they are or where they come from no one really knows. Yet enough documented sightings exist that some things—some creatures—must surely exist. Whether these dark monsters are relics of antiquity or part of a shadowy civilization inhabiting the bowels of the earth is unknown.

What is known: the sight of these unspeakable creatures is enough to drive some men mad.

## 22. Sadistic Surgeons From The Skies

"The Earth is a farm. We are someone else's property." — <u>Charles Hoy Fort</u>

If timeshifts exist—and there is certainly evidence to support their reality—then a very dark side to their nature also exists.

Did you know blood sometimes showers from the sky—human and animal blood? Other things just as disturbing have been cataloged: parts of organs, lumps of flesh, shards of skeletal remains...the list goes on and, frankly, some are too terrible to dwell upon.

Researcher and popular writer-investigator Charles Fort explored many of the oddities of space and time, sometimes publishing compilations of ghastly things that fall out of the sky.

The unsettling phenomena he reported in several worldwide bestselling books are still occurring today. And without a doubt the most frightening phenomenon is the appearance of body parts falling from...somewhere. Seemingly tossed recklessly to the ground like the thoughtless litter of some celestial butcher, the deluge of organic detritus horrifies the witnesses.

### It's Raining Flesh And Blood

An example of these gruesome incidents happened in Sheffield, England during the month of October, 1922.

Without warning, a quarter acre of English pasture was inundated by tons of sliced flesh, blood, entrails, kidneys, hearts, pieces of lungs and mashed brains. Coating some of the grisly debris—later determined to be entirely of human origin—was a mystifying oily substance.

As parts of arms, legs and torsos were also found among the remains, one farmer who witnessed the deluge of death remarked afterward the field resembled a makeshift abattoir.

The organic shower lasted for several minutes and witnessed by dozens of area residents including a local police officer.

No explanation was ever offered, or even attempted.

Investigations were undertaken by local police, forensic experts and Scotland Yard. University professors created absurd theories. The answer, however, may once again be found in a weird permutation of time. Unlike the timeshifts discussed in earlier chapters, debris appearing out of nowhere—and especially tumbling out of a cloudless sky—might have its origins in timestorms.

## Timeshifts And Timestorms

A timestorm differs from a shift. Apparently, whatever slips through the twisting portal of time does not necessarily have to arrive whole on the other end.

Caught in a timestorm, objects, animals, even people can be ripped and torn apart. Some hapless creature, caught in a turbulent shift and transported across space and time may arrive at their destination in less than pristine condition. With the number of incidents that occur, a certain amount of shredding can be expected, and that would account for the blood and guts that occasionally rain down on the astonished heads of the witnesses.

A time "shiftee"—as it would seem from collected evidence—may be transported across continents and time only to re-materialize anywhere: on land or sea, even a mile above or below the earth. Nature does not seem too particular where it deposits the shiftee, nor whether what is being transported is animal, vegetable or mineral.

## Characteristics Of Timestorms

A timestorm usually has one of two striking characteristics:

1. The storm is accompanied by an extremely unusual, bizarre-looking cloud: a timecloud.

Its composition is dissimilar to the normal water-vapor cloud . This cloud is often a precursor to—or manifestation of—a timestorm. It appears dark red in color and is ringed by a band of black. Clouds of this type are described as moving slowly across the sky and then hovering in one spot while disgorging a terrifying payload. Upon precipitating things—living, dead or dying—upon the countryside the cloud drifts away or slowly evaporates.

2. A second characteristic of timestorms is a visible shimmering in the air caused by some type of atmospheric refraction. No cloud appears.

Seen in daylight, the phenomenon looks very much like heat rising off asphalt on a hot summer day. Instead of temperature differentiation causing the refraction, however, it is a ripple in the time space continuum.

Because of the spectacular events that follow the appearance of a timestorm, few witnesses have described the refraction. Researchers of the phenomenon suspect, however, that the atmospheric disturbance accompanies every incident except when the storm appears as a timecloud.

MYSTERIES OF THE MULTIVERSE

# STRANGER THAN FICTION

## 23. Terrifying Giant Five Foot Spiders Spotted In Congo

Spiders might exist that have crawled out of nightmares. They're called the "J'ba FoFi" (giant spider, pronounced ch-bah foo fee) in Central Africa.

Many people might define a giant spider as one that's bigger than their hand. Some may think bigger and envision the horrifying Goliath 'bird eating spider' that dwells in the darker corners of the ancient Amazon rain forest. That eight-legged terror spans a whopping 14-inches.

Unfortunately, those people aren't thinking big enough.

The size of the Congolese Giant spider—when its legs are included—is said to be up to five feet across.

According to cryptozoologists (researchers that investigate unknown creatures that have not been recognized by orthodox science), most of the J'ba FoFi dwell in the Congo. Natives tell stories of the giant web-nests the spiders build, similar to a trap-door spider.

Most of the many anecdotal tales describe the spiders digging a shallow tunnel under tree roots and camouflaging it with a large bed of leaves. Then they create an almost invisible web between their burrow and a nearby tree, booby-trapping the whole thing with a network of trip lines. Some hapless creature—soon to end up on the menu—will trip the line alerting the spider. The victim will be chased into the web. This predatory entrapment is similar to some species of tarantula.

Presumably, the J'ba FoFi eggs are a pale yellow-white and shaped like peanuts. Natives claim the hatchlings are bright yellow with a purple abdomen. Their coloration becomes darker and brown as they mature.

Some of the natives indigenous to the regions in the Congo where the J'ba FoFi has been seen assert that the spider was once quite common, but has become very rare.

Other than the testimonies of natives, the fullest account by Westerners appears in a cryptozoological book by George Eberhart [<u>Mysterious Creatures – A Guide to Cryptozoology</u>]. On page 204, Eberhart relates the terrifying experience of an English couple traveling through a jungle region of what is now called the Congo:

"R.K. Lloyd and his wife were motoring in the Belgium Congo in 1938 when they saw a large object crossing the trail in front of them. At first, they thought it was a cat or monkey, but they soon realized it was a spider with legs [spanning] nearly 3 feet [across]."

Famous naturalist and cryptozoologist, William J. Gibbons, has hunted for what some think may be a living African dinosaur called Mokele-mbembe. On his third expedition in search of the creature he came upon natives who related their experiences with giant spiders. He shared his experience with readers upon his return to Canada:

"On this third expedition to Equatorial Africa, I took the opportunity to inquire if the pygmies knew of such a creature [giant spider], and indeed they did! They speak of the Jba Fofi, which is a "giant" or "great spider." They described a spider that is generally brown in color with a purple abdomen. They grow to quite an enormous size with a leg span of at least five feet. The giant arachnids weave together a lair made of leaves similar in shape to a traditional pygmy hut, and spin a circular web (said to be very strong) between two trees with a strand stretched across a game trail."

This is exactly the same description that other researchers have heard. Although the spider seems to have been spotted mostly in the Congo, there are reports of the same—or similar—spiders inhabiting Uganda and the Central African Republic. "These giant ground-dwelling spiders prey on the diminutive forest antelope, birds, and other small game, and are said to be extremely dangerous, not to mention highly venomous," Gibbons states. "The spiders are said to lay white, peanut-sized eggs in a cluster, and the pygmies give them a wide birth when encountered, but have killed them in the past. The giant spiders were once very common but are now a rare sight."

Many of the natives describe the spiders as once being numerous, but now a vanishing species. Encroachment by civilization in the form of rain forest being converted to farming may have driven the spiders from their natural habitats.

"[Although their numbers are dwindling] they are still encountered from time to time. The Baka chief, Timbo, casually mentioned to us that a giant spider had taken up residence in the forest just behind his village in November 2000, when I and Dave Woetzel from New Hampshire had visited him! He did not think that we would have been interested in the creature as our interest was focused on Mokele-mbembe at the time! Valuable evidence had eluded us."

Cryptozoologists—like any other researchers—sometimes only get the information they specifically ask for.

If these giants do indeed exist, their physiology is puzzling. As some entomologists have rightly pointed out, spiders of that size would have to overcome the limitations of their exoskeletons. In addition to that hurdle, many of the more primitive arachnids have a primitive book-lung respiratory system. Modern spiders, however, often have a trachea and book-lungs. That combination allows for a smaller heart, more efficient blood flow and greater speed and stamina. If the Congolese giant spiders exist, they would most likely have both trachea and book-lungs.

"On questioning our group of six Baka guides," Gibbons narrative continues, "they have all seen these spiders at one time or another and state that they are quite capable of killing a human being. According to the Baka (and the Bantu hunters who have encountered them) the giant spiders were once surprisingly common and would often construct their lairs very close to human villages. They have become quite rare now thanks mainly to the deforestation of Central Africa, but my guess would be that they are still to be found in numbers in the vast and still untouched forests of the former [Belgian Congo or Zaire] where the Lloyds encountered one in 1938."

Gibbons knew the Lloyds personally and adds that Mr. Lloyd tried to get a photo of the spider while Mrs. Lloyd was so stricken with fear all she wanted to do was return to their home in Rhodesia.

Other stories of giant spiders abound. Some of the stories are little more than spotty tales told in the villages of unnamed missionaries whose porters were killed by giant spiders.

An English missionary named Arthur Simes related an incident that occurred in Uganda during the 1890s. While trekking near the shore of Lake Nyasa, his porters became entangled in a monstrous web. Several giant spiders swiftly descended upon them, injecting the men with poisonous venom. Later, all their extremities swelled, they grew feverish, delirious and then died.

Simes claimed he drove the giant spiders off with his pistol.

Whether the Congo spider is real, or a myth remains to be seen. And hopefully, whomever the researcher is hunting for it, he will see the spider before it sees him.

## 24. The Mysterious Monster Of The Amazon

The world is filled with secrets. Many of them it does not give up easily. Empires have risen and empires have fallen in the midst of mysteries-riddles that remained unsolved through the centuries.

One such mystery has haunted the darker, steamier regions of the fabled land of the Amazon. The Amazon: that dark, forbidding river sluggishly twisting its way through Brazil and eight other South American countries. It is a river second in size only to the mighty Nile and it's so wide that no bridge passes over it.

As far back as the Aztecs, legends have been spoken about a monster snake. The Aztecs of central Mexico made it one of their most powerful gods: Quetzalcoatl.

In the centuries that followed , the indigenous peoples of the Amazon often spoke of the Yacumama—the snake of the water. European and American herpetologists shrugged off the talk as myths or as references to the great aquatic boa, the anaconda.

According to the indigenous peoples other giant snakes inhabited the Amazon's shadowy realm too: the Sachamama and the Minhocão, a snake that some Amazon natives believed could alter the land as it passed through.

Despite the herpetologists' belief, the natives were not speaking of big anaconda by different names. They spoke of true monsters-leviathans so huge that the anaconda would be small in comparison. The snakes the natives sometimes spoke of —in fear and awe over village campfires and in the safety of their homes—measured 120, sometimes 160 feet long. The heads of these mammoth creatures were said to reach six feet wide. They could shoot down prey with explosive jets of water, topple trees in their passage and change the course of minor tributaries.

During the year 1906 the world-famous explorer Major Percy H. Fawcett claimed of encountering a gigantic anaconda while traveling up the Amazon River. He shot the creature and observed it as it lay dying.

He recalled: "We stepped ashore and approached the reptile with caution. It was out of action, but shivers ran up and down the body like puffs of wind on a mountain tarn. As far as it was possible to measure, a length of 45 feet lay out of the water, and 17 feet in it, making a total length of 62 feet...such large specimens as this may not be common, but the trails in the swamps reach a width of six feet and support the statements of Indians and rubber pickers that the anaconda sometimes reaches an incredible size, altogether dwarfing the one shot by me. The Brazilian Boundary Commission told me of one killed in the Rio Paraguay exceeding 80 feet in length!"

Yet the professional academicians and peer-reviewed herpetologists were far from convinced. Monstrous snakes just seemed so patently outlandish.

So the controversy festered for another century until two brothers, Mike and Greg Warner, mounted an expedition into the Amazon jungles hunting for evidence of the monster snakes. The expedition was inconclusive, although they recorded mammoth snake trails and took testimonies from natives who claimed to have seen the Yacumama.

Mike Warner talked to hundreds of natives and workers who had encounters with the Yacumama. He researched thousands more. He notes that both native tribes of certain African regions and natives near the Amazon River in South America describe a huge snake that "carries its water with it." Although the first expedition failed to find the elusive Yacumama, the brothers were undeterred. After raising new funds they mounted another expedition to the Amazon.

It seems that they found what they were looking for. In fact, their findings are so credible that the National Geographic Society expressed serious interest in the brothers' research and findings.

What they found tallied with previous eyewitness reports.

Over the years, many sightings of Yacumama have included descriptions of the snake as sprouting horns on its head. This peculiar feature, mentioned in so many reports coming from independent observers up and down the Amazon, has led Warner to his hypothesis that the Yacumama could be a prehistoric version of the modern day caecilian. Most of the 50 or so species of caecilian that are cataloged do have a groove running along either side of the head that contains retractable tentacles. To untrained observers, they may appear as horns.

According to Mike Warner, "The exact species of this creature is unknown but we believe that the physical characteristics and behavior are that of a snake [or amphibian] with behavior similar to a caecilian." [An amphibian creature similar to a snake.]

Most of the witnesses who have sighted a Yacumama did not spend too much time studying the creature —they usually happened upon it by chance and then turned tail and ran for their lives.

Warner's research led him to discover that the Yacumama seeks its prey near regions where two rivers merge into one, called a "confluence." He determined that those areas provide the mammoth predators a constant supply of food.

He hypothesizes that such a capability may have one or more of the following purposes:

Stunning prey or dislodging them out trees. (The Yacumama reportedly engorges water and shoots it at its prey like a water cannon.

As it "carries its water with it" it is possible that it may use this water pressure to support its skeletal structure as it moves through the jungle.

It can also use water as an instrument for burrowing [into the ground like a worm does, hence the similarity in some traits to a caecilian].

## 25. Legendary Ghost Camels Of The American West

The old timers know about the ghost camel.

Some of those grizzled desert rats that roam the towns and byways off Interstate 41 swear they've seen it. Their memories are haunted by the thing known only as the Red Ghost—a camel that haunts the desert.

Yet Highway 41 is not the only road girdled with galloping ghosts.

Other roads—some desolate and unnamed—wind their way through lonely sand swept vistas across the bleak desert Southwest. Those roads skirt ramshackle towns that appear on the state road maps of west Texas, New Mexico, Arizona, Nevada and Utah as flyspecks and afterthoughts. Lingering near sun-weathered shacks (that preposterously claim to be service stations) other sun-baked folk often gather to trade personal tales of the desert and some of its more obscure legends.

Yes, those old timers also know. Some of them swear that the Red Ghost is real; they've seen its spectral form and its ghastly, headless rider with their own timeworn eyes.

The legend of the Red Ghost is rooted to the U. S. Cavalry's disastrous experiment with camels. Promoted in 1855 by then Secretary of War, Jefferson Davis, $30,000 was raised by Congress to deploy Middle Eastern camels in a scheme to keep supplies flowing along the southwestern state's wagon train trails.

On paper the idea seemed a stroke of genius; in practice it became an embarrassment wrapped in a boondoggle.

While the U.S. Army's failed flirtation with camels is a fairly well-known story, other herds of camels were imported into America about the same time. During 1858 and 1859 small herds of camels were imported into Texas. Most of the camels were either shot or escaped into the desert.

The very next year, 1860, more camels arrived. These beasts were shipped into San Francisco. The idea was to employ the camels in a makeshift caravan to supply the miners working Nevada's silver industry. Yet another group were brought into Babylon by the Bay in 1862. Those never made it into Nevada, however, as they were hurriedly sold off to interests in British Columbia.

As time progressed the camels became superfluous. After the Civil War the railroads expanded across the desert terrain and camels became much more of a nuisance than an asset. For example, the mining companies in Nevada eventually released most of their camels into the desert wilderness. The miners had their fill of the camels' notoriously bad temperament: they tended to bite, kick and spit at humans.

Eventually all the camels imported from 1855 to 1862 either died, were killed off, escaped or were released. Those that escaped or were released provided the grist for the ever-turning mill of camel folklore. Indeed, many experts contend that claims of living camels roaming the Great American Southwest is nothing but fanciful folklore—lurid stories cut from whole cloth. After all, they ask, how could pockets of camels survive until today, reproduce in the wild, and find enough water to live? It certainly seems impossible, they scoff!

Impossible, except for the bothersome fact that so many eyewitnesses have sworn steadfastly over the last one hundred twenty years that they have had camel encounters. Decade after decade persistent reports emerge claiming that solitary or groups of these displaced animals continue to wander at will through the more arid regions of the American deserts. Among these reported sightings are those of the fabled ghost camels.

The ghost camel stories are legion. Of all the stories perhaps the most colorful is that of the Red Ghost.

According to the old timers a young army recruit assigned to ride one of the stubborn beasts had a great deal of difficulty controlling it. Exasperating matters was the fact that the young man feared the camel.

The commander of the unit decided to teach the recruit a lesson and help him overcome his fear of the animal. Therefore, he ordered some of the other men to seat the recruit upon the camel's back and securely tie him in place. The theory was that the camel would be unable to toss the recruit off.

Laughing, several men gave the camel a few good smacks on the rear and the beast stampeded off, frightened. They gave chase but soon lost it in the desert. The camel, nicknamed the "Red Ghost" because of its rust-colored fur, was not seen again. The poor soldier lashed to its back was also lost amidst the sun-baked sands.

During the ensuing years sporadic tales drifted throughout California, Nevada and Arizona that a red camel was sometimes spotted well off the established trails. Upon the camel's back rode the emaciated figure of a man.

Little detail accompanied these tales until the year 1883 when a woman's body was tragically found trampled to death in the desert. Investigation of the corpse indicated it had been attacked by a large animal with reddish fur. Nearby townsfolk immediately suspected the Red Ghost for what else could be marauding through the desert with red fur?

Not long after the grisly discovery large hoof prints were found in the same area. Then a nearby rancher came forward with an eyewitness sighting. He claimed he saw a red camel with a rider. Others reported sightings to the local newspapers. Finally several desert prospectors laid claim to spotting the Red Ghost and seeing something fall off its back...apparently a human skull.

Over the next ten years the Red Ghost terrorized the desert community. Then one day in 1893 an Arizona farmer declared he had killed the Red Ghost. Of the headless rider no sign could be found, but the terrible leather straps that had bound the hapless soldier to the crazed camel were still hanging from the great beast's back.

Despite the killing, reports continued to be made of encounters with the Red Ghost. The reports have continued right into the 21st Century.

From time to time breathless people call their local radio station or sheriff's office to report a mad, red camel bounding through the desert with long white bones strapped to its sides—all that remains of the creature's dead rider. So if you happen to be riding any of the lonely byways off Interstate 41 in the Great Southwest someday and spot alongside the road a great, red spitting camel with fire in its eyes don't question your vision or your sanity. You'll just be the latest witness to the ongoing saga of the wandering Red Ghost.

## MYSTERIES OF THE MULTIVERSE

MYSTERIES OF THE MULTIVERSE

# APPENDICES

## Appendix I. The Quantum Brain, Parallel Worlds And Timeshifts

Modern physics has determined that energy is information. Therefore entropy is a condition of increasing chaos in the order of the underlying information contained in quantum energy.

The information contained in energy is universal. This has been shown by multiple experiments at the quantum level. These experiments have shown that a quantum particle or wave that has been excited in one region of space affects another particle or wave in a completely different region of space. Distance does not matter—the reaction of one quantum particle/wave to another occurs whether the other particle/wave is several hundred feet away or 100 billion light years away.

It has also been documented that by simply observing an object at the quantum level, it "forces" the quantum object to "make a choice."

Quantum can either be expressed as a particle (the basic building block of matter) or a wave (the basic building block of energy). Astoundingly, the act of a conscious intelligence directly interacts with the basic foundation of reality and influences that reality in measurable ways. This has been known to be true for some decades. It has been demonstrated in laboratory experiments, so it is not theoretical, but a measurable phenomenon.

Taking that documented effect into consideration, it leads us to the realization that conscious intelligence itself is inextricably linked to quantum reality and cannot only be influenced by the quantum matrix of reality, but can in itself directly influence the quantum matrix of reality and by extension, thus has the ability to modify the underlying reality of existence.

It has been demonstrated repeatedly during the past several years that the matrix of our universe (the physical, observable reality we are a part of) is linked to a multiple of other "realities" each of which exists independent in and of itself but can also interact with other realities within the matrix of a multiverse.

Supporting this mathematical evidence, laboratory experiments have measured a field of ubiquitous energy called "Zero Point Energy." Data suggest this energy has been tapped not from our universe, but from "somewhere else."

Experimental evidence for the existence of a multiverse can also be seen with the manipulation of elementary particles within the famous "Josephson's junction" experiments wherein matter and energy transmitted between the junctions simply "winked out of existence" only to reappear at the other side of the junction without observably crossing the physical space between.

Other experiments, such as the groundbreaking experiments involving the transmission of information in the form of energy between one point in physical space to another - instantaneously - demonstrated that the theoretical limits of the speed of light can be circumvented and an underlying physical reality that neither Newton's or Einstein's views of the universe (and their attendant physical laws) take into consideration.

For instance, certain observations within particle accelerators at the Fermi Lab (and others) have recorded anomalies that can only be accounted for by interaction with some other reality (or co-existing dimension). [Quantum mathematics, quantum physics and cosmology all tend to support this.]

Consciousness, which cannot be accounted for strictly by the bio-physicality of the human brain, has been demonstrated to be holistic in nature. There have also been experiments that tend to reveal that consciousness is not only holistic in its ability to gather data from the available senses, distill that data into an ordered hierarchy of information, and then manipulate that information into a condensation of completely new information (the creative process of consciousness), but can actually manipulate (or be manipulated by) the interaction of the surrounding environment not only at the macro level but to the very matrix of the quanta itself.

All of this information lends credence to the idea that everything that composes the known universe and the unknown multiverses, reduced to its most elementary level, is nothing more than a constant flow of information that can be interacted with by itself, other outside sources of information (one or more of the multiverses), or any conscious entity that can observe any of the informational processes of one or more of the existing realities.

Such a scenario then also lends some credence to the controversial hypothesis that the phenomenon termed "space-time" does not actually exist and has no more substance than a shadow does to the interaction of macro-matter blocking protons emanating from an independent light source, (those photons being inhibited from streaming unfettered onto a background).

Therefore, everything is reduced to simple units of information: quantum matter, quantum energy, quantum time, quantum space—all is nothing more than forms of an informational matrix , a matrix of which consciousness itself is the simultaneous decoder, re-assembler, re-arranger, distiller and manipulator…and in some instances, creator.

# MYSTERIES OF THE MULTIVERSE

For consciousness to have such properties and abilities, consciousness itself must be part of the quantum matrix—a quantum matrix that is not confined to one reality but spans all realities (all universes) simultaneously. If this is valid, then every quantum potential is only awaiting stimulation with an interaction by a quantum consciousness to realize a modified nature of information.

Have observations been made supporting this contention?

Yes!

The newest observational physics has deduced that every observation we make of our observable universe is measurably changing the information of existence at the quantum level and introducing increased chaos (entropy) to this reality thus in effect hastening its eventual demise. "Demise" would be defined as "a catastrophic loss of order within the information supplanted by absolute chaos." To draw an analogy, it is very similar to the corruption of information in a computer program, or just another way to describe non-existence.

This breakthrough paves the way for a new physics that could account for the basic building blocks of quantum information and their subsets, mainly: space, time, gravity, the weak force, the strong force, entropy and even so-called paranormal forces or events that have been documented throughout history such as teleportation, telekinesis, telepathy, precognition, significant time anomalies, UFOs and other phenomenon that have largely been dismissed because they did not fit into the accepted models of physics, cosmology or the perceived nature of reality.

The holonomic, quantum brain interacts with the universe at the quantum level. Therefore, cognitive awareness (the consciousness) can be shaped by contact with existence in this universe and all the facets of the multiverse) can be shaped and directed by consciousness.

This concept was the basis for the speculative science underlying the mind boggling plot of the landmark science fiction film, "Forbidden Planet."

On that planet, Altair IV, a race of super beings called the Krell established an advanced civilization without instrumentalities. In other words they were able to manipulate energy and matter simply through an effort of conscious will and create anything sending it anywhere on the planet for any purpose. They had advanced beyond their dependency on machines.

Hollywood speculation has become hard science in the 21st Century laboratory. It's been established that cognitive awareness has a measurable effect upon quantum particles and waves.

Granted that today's breakthroughs into the nature of the symbiotic relationship between an aware universe and aware intelligences is a far cry from establishing a Krell-like civilization, but the key to accelerating the advancement of research in this field is, appropriately enough, the creation and utilization of quantum computers.

This is being done even now.

In summation, consciousness exists in a multiverse that's composed of an unknown number of dimensions—or realities—and the observable universe is only a part of a greater reality.

Consciousness is also multidimensional and transverses the multiverse but only on extremely subtle quantum levels.

Therefore, each of consciousness awareness is continuously acting, reacting and interacting with the quantum matrix and by its very nature transverses all realities. This state is one that's normally invisible.

Therefore, the nature of time, of space, of energy and of matter are all governed by the underlying matrix of quantum information and to a degree—a degree yet undetermined and unknown, and only partially understood—the nature and purpose of consciousness is primarily to control, manage, manipulate and influence the reality of the dynamic multiverse.

## Appendix II. Quantum Entanglement And The 'Time Barrier'

Researchers have come across an odd phenomenon while studying the properties of quantum physics. Again and again experiments have shown that two linked particles seem to be entangled and what happens to one happens to the other. Surprisingly it doesn't matter if they're one mile apart or 100 miles apart. Theoretically the entanglement might hold even if they were 100 million light years apart.

What can possibly tie the two particles together that they violate all the rules and just ignore distance and—more importantly—time?

Perhaps the answer has arrived—in the nick of time—from two physicists unraveling the weird state of entanglement. Their solution, if testable, may accomplish more than solving a thought problem. It might provide the path towards crunching information in quantum computers and reveal a bit more of how the multiverse all fits together. The lead author of the study attempting to untangle entanglement is quantum physicist S. Jay Olson of Australia's University of Queensland. In an interview with wired.com, Olson explained how particles can interact outside of time and space. "You can send your quantum state into the future without traversing the middle time."

## Using The Quantum To Circumvent Time

The universe's speed limit is the speed of light; 186,000 miles per second. But for entangled particles operating in the quantum speed limits don't exist. Entangled particles are linked outside of the space time continuum.

It's believed that so-called "ordinary" entanglement binds two atomic particles—normally electrons or photons—so closely that they literally share the same quantum state. Disturbing one particle elicits an instant change in the other no matter where that other is in space.

This amazing property allows physicists to utilize entangled particles to transmit information and encrypt data with unbreakable coding. The property of entanglement also allows the creation of super computers so fast that 2011 computing technology would be relegated to the Stone Age.

But entanglement offers the promise of even more: scientists believe using its inherent properties in certain ways can facilitate data burst transmissions of massive amounts of information anywhere in the universe using a special protocol: quantum teleportation.

Using quantum teleportation to send information only employs the use of a handful of atoms.

This is Star Trek technology. Quantum teleportation for the transmission of information is similar to the sub-space communication envisioned by Gene Roddenberry. It's also the technology that Captain Kirk used to save his own skin on several occasions.

## Future's Past

Olson and his Queensland colleague Timothy Ralph have posted a mathematical model on the website arXiv.org demonstrating how quantum information can be sent anywhere the system is set up and from the past into the future.

"If you use our timelike entanglement, you find that [a quantum message] moves in time, while skipping over the intermediate points," Olson told wired.com. "There really is no difference mathematically. Whatever you can do with ordinary entanglement, you should be able to do with timelike entanglement."

Other experiments by other researchers have shown how quantum physics can reverse the laws of causality. Normally a cause is followed by an effect. But some experiments have revealed that entangled particles can respond to a cause before the cause has occurred.

Using that principle, it's logical to assume that information can not only be sent forward in time, but backwards in time as well.

Olson and Ralph, however, do not address reverse causality in their study.

## RESOURCES AND LINKS FOR APPENDIX I, "THE QUANTUM BRAIN, PARALLEL WORLDS AND TIMESHIFTS"

Web Links and PDF White Papers

BASIC LOGIC AND QUANTUM ENTANGLEMENT P. A. Zizzi
Dipartimento di Matematica Pura ed Applicata
Via Trieste, 63-35121 Padova, Italy zizzi@math.unipd.it
http://www.quantumbionet.org/admin/files/0611119.pdf

Holonomic Brain Theory - Mohsen 05-13-2005
http://www.toequest.com/forum/neuroscience-articles/ 445-holonomic-brain-theory.html#post2473

Holonomoc Brain Theory
Dr. Karl Pribram, Georgetown University, Washington, D.C.
http://www.scholarpedia.org/article/Holonomic_brain_ theory

Science and the Mind ~ Audio Slideshow Presentation
Dr. Roger Penrose, Oxford University
http://online.kitp.ucsb.edu/plecture/penrose/

Time-shift attack in practical quantum cryptosystems
Authors: Bing Qi, Chi-Hang Fred Fung, Hoi-Kwong Lo, Xiongfeng Ma (Center for Quantum Information and Quantum Control, University of Toronto)
http://arxiv.org/abs/quant-ph/0512080v3

Emergent Consciousness: From the Early Universe to Our Mind
Authors: P. A. Zizzi
http://arxiv.org/abs/gr-qc/0007006

Is Quantum Mechanics Controlling Your Thoughts?
"Science's weirdest realm may be responsible for . . . consciousness itself." - by Mark Anderson

Discover - Physics & Math / Subatomic Particles
http://discovermagazine.com/2009/feb/13-is-quantum-m echanics-controlling-your-thoughts
   "Funda-Mentality" Is the Conscious Mind Subtly Linked to a Basic Level of the Universe? ~ Stuart Hameroff, M.D.
Arizona Health Sciences Center Tucson, AZ 85724
http://www.quantumconsciousness.org/penrose-hameroff / fundamentality.html

THE HOLOGRAPHIC BRAIN ~ With KARL PRIBRAM, Ph.D.
From "THINKING ALLOWED" Conversations On The Leading Edge Of Knowledge and Discovery With Dr. Jeffrey Mishlove
http://twm.co.nz/pribram.htm

Quantum Time Revisited
http://www.awitness.org/unified/pages/page2/quantum_ time.html

Quantum Biosystems
http://quantum.ibiocat.eu/eng/index.php?pagina=81

Schrdinger's Cat Thought Experiment
http://www.answers.com/topic/schr-dinger-s-cat

Discussion: Quantum brain dynamics
http://en.wikipedia.org/wiki/Talk:Quantum_brain_dyna mics

The Creation of Reality or Dynamic Consciousness and the Reality of Free Will
http://free-will.de/

# MYSTERIES OF THE MULTIVERSE

The Quantum Brain
Bose-Einstein condensates
Quantum Brain Dynamics
Microtubules and the mind
Stapp's quantum paradigm
cf. Quantum computation
http://nonlocal.com/hbar/qbrain.html

TIME, CONSCIOUSNESS AND QUANTUM EVENTS IN FUNDAMENTAL SPACETIME GEOMETRY ~ Stuart Hameroff
http://www.quantumconsciousness.org/Time.htm

Time shift phenomena in Einstein Rosen solitary-like waves
A D Dagotto, R J Gleiser and C O Nicasio
Facultad de Matematica, Astron. y Fisica, Univ. Nacional de Cordoba, Argentina
http://www.iop.org/EJ/abstract/0264-9381/8/6/015

Consciousness, Whitehead and quantum computation in the brain: Panprotopsychism meets the physics of fundamental spacetime geometry
Stuart Hameroff
http://www.quantumconsciousness.org/Whitehead.htm

Video presentation: Neural Correlate of Consciousness
David Chalmers
http://www.youtube.com/watch?v=Yj1JaidwGvM&feature=PlayList&p=708D5102535C25F1&playnext=1&playnext_from=PL&index=1

Video Lecture: Toward Science of Consciousness "The Contingent Brain"
Ken Mogi at Hong Kong Polytechnic University, 2009
http://www.youtube.com/watch?v=6cxT20PYPY0

Scientific White Papers in PDF Format

How Consciousness Creates Reality ~ Claus Janew
http://free-will.de/reality.pdf

Consciousness, Neurobiology and Quantum Mechanics: The Case For A Connection ~ Stuart Hameroff
http://www.quantumconsciousness.org/documents/case_0 00.pdf

The Brain Is Both Neurocomputer and Quantum Computer ~ Stuart R. Hameroff
http://www.quantumconsciousness.org/documents/CogSci pub.pdf

Consciousness, the Brain, and Spacetime Geometry ~ Stuart R. Hameroff
http://www.quantumconsciousness.org/penrose-hameroff /cajal.pdf

BASIC LOGIC AND QUANTUM ENTANGLEMENT
http://www.quantumbionet.org/admin/files/0611119.pdf

QUANTUM PHASE SHIFT CONSIDEREDAS A GR-LIKE TIME DILATION ~ HOWARD A. LANDMAN
http://www.riverrock.org/~howard/QuantumTime4.pdf

MYSTERIES OF THE MULTIVERSE

# ABOUT THE AUTHOR

### TERRENCE AYM - AYMING HIGHER

"The most important person in my writing is YOU. And I try never to forget that and strive to make what I write entertaining and worth your time." ~ Terrence Aym

"Eclectic, dramatic, exciting..." These are some of the words readers of Aym's works use to describe his writing.
    Once during a radio interview, Terrence Aym was asked what motivated him to write. He responded that he writes for two primary reasons: the first is to entertain and inform his readers; the second, writing gives him personal pleasure.
    Aym has appeared as a guest on talk radio across the US and internationally. His articles have been read, discussed and sometimes heatedly debated by millions.
    The news media has also discussed and debated Aym's articles and ideas. Among those that have featured articles about Aym and his writing are ABC News, FOX News, TIME, Business Insider, Smithsonian, Nature, National Review, Discover Magazine, Pravda, Salem-News.com, The Nation, and many more.

    Several of Aym's internationally viral articles have been re-published in Spanish, French, German, Portuguese, Korean, Russian and Chinese.
    One of his science articles made the syllabus of an LSU course.
    Thousands of bloggers follow Aym's articles.
    Aym is currently working on several more books.

Following Aym on Twitter? If not, do it now! - http://twitter.com/TerrenceAym

Once you're signed up you'll get the latest cutting edge articles with a different perspective than all the rest. Fun, thought-provoking, informative...sometimes controversial. But never boring.

Terrence Aym also invites you to join him on his public Facebook page, Write On!

MYSTERIES OF THE MULTIVERSE

Published and distributed by Before It's News, Inc.
Visit us at www.beforeitsnews.com.

Made in the USA
Lexington, KY
29 February 2012